Frank Sherman Peer

Soiling, Summer and Winter, or, Economy in Feeding Farm Stock

Relating the experience of the author, and giving the latest and most economical

methods of summer and winter feeding by the system of soiling

Frank Sherman Peer

Soiling, Summer and Winter, or, Economy in Feeding Farm Stock
Relating the experience of the author, and giving the latest and most economical methods of summer and winter feeding by the system of soiling

ISBN/EAN: 9783337256845

Printed in Europe, USA, Canada, Australia, Japan

Cover: Foto ©berggeist007 / pixelio.de

More available books at **www.hansebooks.com**

SOILING;

SUMMER AND WINTER,

— OR —

Economy in Feeding Farm Stock.

Relating the experience of the Author, and giving the latest and most economical methods of Summer and Winter Feeding by the system of Soiling.

— BY —

F. S. PEER,
EAST PALMYRA, N. Y.

PUBLISHED BY THE AUTHOR.

ROCHESTER, N. Y.:
HERVEY H. SMITH, PRINTER, 117 AND 119 WEST MAIN STREET.
— 1882. —

DEDICATION.

To the farmers' sons of America this book is dedicated, with the best wishes of the Author, and in the hope that they may find within its pages both the inclination and the encouragement to pursue agriculture as a business, instead of leaving the farm for some so-called "higher" profession.

INTRODUCTION.

TO the intelligent farmer, who keeps abreast of the times by the perusal of one or more good agricultural papers, there is little need of an introduction to the subject of soiling. He is already familiar with the subject and requires no formal presentation of what this book contains. To any others who may chance to peruse these pages, I will say that the work is designed to answer the perplexing questions, how can a farmer enrich his soil in a sure and economical way; how improve and increase his farm stock; how supply them with the most nutritious food at the least cost; how obtain a full flow of milk from his cows during the entire season independently of parched pastures or drouths; how increase his acreage without buying more land. An attempted solution of these and kindred questions, will be found in the following pages.

In relating my own experience in conducting the soiling system, it is not my purpose to boast of what *I* have done, or what *I* can do. Any other farmer may practice the system with the same or even better results; for my knowledge of it is not perfect, and each year's experience reveals many new advantages of the system.

INTRODUCTION.

In these days of hurry and push a man of business has not time to waste in reading refined and imaginary theories, but he seeks how he may shun the failures of others, how he may profit by the experience which has *led the actual worker to success*; thus avoiding the hard and often discouraging teachings of experience. I have, therefore, avoided speculation, and endeavored to present this subject in the light of "vulgar truth," giving only what I have found, after repeated trials, failures and successes, to be the most economical and the best methods of condcuting the soiling system. I do not pretend to say that my conclusions will be found infallible under all circumstances, but I hope that by showing how the general principles of the system have been adapted to my circumstances, the reader may obtain a clear view of its workings and be enabled to adapt it with such alterations as the different conditions under which he may operate shall suggest.

I am not farming for pleasure, although I find a great deal of pleasure in it. I follow farming as a *business*, in gaining my daily bread, the acquisition of which depends upon my own endeavors, and the blessings of Providence. I am not backed by a profitable business in the city to aid in carrying out mere ideas or visionary theories, regardless of expense. My readers need not fear, then, that they are about to be entertained by the diversions of a "fancy farmer."

I have no apology to offer for presenting this subject to the public in book form. I humbly acknowledge that it is not done "at the earnest solicitation of numerous friends," but because I am intensely interested in farming as a business or profession, and would fain see more of our intelligent young men engage in this pursuit. And to such of my co-laborers as may wish to know them, I shall explain in detail how I found the answers to the different questions which confront and perplex us in the beginning of our own life work.

As a literary writer, I make no pretensions. If this work is well received, it must be entirely on its merits as a record of the personal practical views and experience of a farmer. And if the reader finds as much pleasure in perusing these pages as it has afforded me to write them, I shall feel that my labor has not been spent for naught, nor his attention engaged in vain.

SOILING.

THE INFLUENCE OF FOOD ON THE DEVELOPMENT OF FARM STOCK.

As the quality and quantity of forage crops depend upon the fertility of the soil, in like manner does the condition of our domestic animals depend upon the quality and quantity of the food which the soil produces. The cow is but a machine for the production of milk, butter, cheese or beef; the sheep is but a maker of wool and mutton; the horse is but a motive power for draft or speed. In either case the raw material employed is simply the food each consumes, and the profit realized by the owner is produced by the amount of food consumed above that required to supply the demands of the physical organization, as the profits of a factory are found in its products after deducting the cost, and as the profit of a steam engine lies only in the power it can exert after sustaining its own motion. "I do not mean to say," says Mr. M. Allender, in a paper read before the London Farmers' Club, " that a cow is like a steam boiler, that the more coal (food) you throw into the furnace the better results you obtain, but I do maintain that the food, both in kind and quantity, has much to do with the important item of profit and loss."

It may be said that forced feeding of farm stock is not only injurious to the animal, but also defeats the end in view. True, but no more so than of the factory or the steam engine. They, also, have limits beyond which they cannot be forced without great risk of lost to the owner.

Says Mr. Hoxie: "It is generally supposed that varieties of domestic animals have sprung from comparatively few families; that all varieties of the horse, from the draft-horse to the Shetland pony, have sprung from one family; that all varieties of the cow, from the little Kerry to the magnificent Short Horn, are from the same original stock." And history informs us that the American trotter, the English hunter, the Clydesdale of Scotland, and the Percheron of France, have a common ancestor in the Arabian horse. What, we may naturally inquire, is the cause of the wide differences between them? If our cows are from the same family, why is it that the Jersey cannot compete with the Hereford for beef? And in the case of horses, why can not the Clydesdale compete with the Hambletonian on the track, or enter the race for the Derby?

These variations in domestic animals have occurred principally within the memory of man. Therefore, the oft repeated adage that "like begets like" cannot be strictly true. We must look further; for, to accept one and not regard the other, would, to say the least, result in uncertainty; but to reject both sends us still further back into ignorance.

Referring again to history, we read: "The *fer-*

tility of the soil of the rich valley of the Tees originated the English Shorthorn ; the barren mountain regions of the northwest part of Ireland, the little Kerry; the fertile plains of western Europe, several varieties of the great draft horse ; and the bleak, barren islands off the north coast of Scotland, the Shetland pony."

From this we may infer that, while "like begets like," the different families of domestic animals may be so improved that, after a few generations, there can be found few points of close resemblance to the original. I shall try to show how this variation, and the development of new and valuable characteristics, are largely due to the *influence of food*, and how a correct knowledge of feeding places in the hands of the farmer a lever power, the force of which cannot be estimated.

The principles of feeding and selection in the process of breeding are no longer experimental, but have become resolved into a science quite exact. Prof. Miles, in his valuable work on "Stock Breeding," says: "The Kerry cattle are a small and hardy race. The scarce supply of coarse food obtained upon their native hills (Ireland) by industrious efforts give a slow growth and a late development of the organization, so that heifers do not breed until six years old." He adds, "Animals of the same breed raised in Massachusetts, under more favorable conditions for development, are larger than the original type and mature earlier, the heifers breeding at the age of three years." He attributes the change to the influence of better food and shelter.

"Long-wooled sheep," says Mr. Stewart, "are properly the natives of the rich lowlands of England which are productive of abundant succulent and nutritious pasture." As in the animal kingdom there is a marked difference between the animals of the same breed which feed upon the barren hills and those which graze in the fertile valleys of the same country, there is a similar variation from a similar cause noted in the vegetable world. In the tropics the forest trees and plants never cease to grow, feeding upon rich plant food during the entire year, thereby developing grand proportions. But as we approach the north and the cold increases, vegetation diminishes, until we reach the Frigid Zone, where, says Dr. Hall, " you can cover a forest of a thousand trees with your hat." As already stated, plants, like animals, live, feed, grow and die, and the development of each depends upon the amount and quality of their respective foods. As vegetation thrives best in the warmer climates, we should learn the importance of providing warm and comfortable quarters for our farm stock during the winter. And, again, as plants thrive best where they obtain their food uniform in quality and quantity during the entire year, the best results in our stock can be reached only by a similar provision for them.

The soiling system, *summer and winter*, offers the very best means known, at the present day, of accomplishing these results. The field of one farmer may produce bountifully while that of his neighbor over the fence may not produce enough to pay the expense of growing and harvesting. Of the former, we may say the land is rich; of the latter, poor. What we really mean to say is that the soil of the one has been well supplied with plant food, while the other has been starved and is hungry. There are thousands of farm animals that do not pay the interest on their purchase price above the cost of their keeping, simply because they are not fed enough. Says Francis Morris, in the Agricultural Report of 1863, "Take two Durham calves from cows of equal purity, sired by the same bull, send one into 'green fields and pastures new,' stable it in cold weather, give it all it can need for its health and comfort; turn the other out to pick its own living on a rocky, barren field, to shiver with cold, to search wearily for food. At the end of a year, compare them. The first will be a fat, sturdy, handsome fellow, sleek, bright, erect of head, courageous and intelligent ; the other, a miserable, melancholy runt, without pluck or beauty, lean, small, and showing scarcely a trace of beauty." He adds, in substance, "Now try to fatten them. The one that was well treated will thrive rapidly; the other, though it may improve, will never attain its proper size, nor be anything but a runt; nor will it be fit for breeding purposes ; the probabilities are that the runt qualities acquired by hard usage will be transmitted."

The influence of food has also a marked effect on the reproductive powers of animals. Says Prof. Tanner Miles, in his work on "Stock Breeding," "When the fall rains have been small, and the herbage more than usually parched, we find unusual difficulty in getting ordinary farm stock to breed. A dry dietary is very unfavorable for breeding animals and very much retards successful impregnation. On the other hand, rich, juicy and succulent vegetation is very generally favorable to breeding." Taking advantage of this, breeders of long wool and mutton sheep manage to have a fresh pasture in which to turn their ewes during the coupling season, or supply them with other green forage, such as tares and rape, sometimes adding grain ;

by so doing the ewes are more apt to drop twins. It is evident that the influence of food has much to do not only with the improvement and early maturity of our farm stock, but also with the question of profit and loss while keeping them. The economy in feeding farm stock in order to attain the greatest improvement, earliest maturity, and largest profits, is the result of continued liberal feeding of the most succulent and nutritious foods abundant in albumenoids, carbo-hydrates, and fat. The economy in feeding farm crops, in order to obtain the greatest growth, the earliest maturity, or the largest profit, will likewise be found in the most liberal supply of plant food, abundant in nitrogen, phosphoric acid, and potash.

He who would practice true economy in feeding should have lettered upon the walls of his stables, and distinctly and indelibly stamped upon the tablets of his memory, the ancient and wise saying, "*Withholding doth not enrich thee nor giving impoverish.*"

The effect of food and selection in breeding, which has already been noticed, has produced, in each family of domestic animals, three distinct characters. Thus, among cattle, are those peculiarly developed for producing beef, milk, or butter; of sheep, are those best for fine wool, long wool, or mutton, and of horses, are those excellent for draft, road, or speed. We may well consider, for instance, how the influence of food has produced in the bovine race one breed excellent for beef, another for milk, and another for butter. The successful plan for improving stock has always been by the perfection of some one valuable quality and making all others subordinate. Says Prof. Miles, "When the entire energies of the system are acting in a particular direction, as they must do to ensure the highest developement of a single quality, there is no residuum of force for the developement of other qualities that are not strictly correlated with the one made dominant." In these days of close competition, it is very essential that the breeder, who would successfully compete with his neighbor, should select a breed (at least the sire) whose ancestry excels in the qualities best adapted to the condition of his farm and the demands of his nearest market. From such animals only will he receive the most profitable returns from his investment. A cow, for instance, which is a "jack at all trades, is master of none," and can not compete with trained milkers, butter makers, or beef producers. Every farmer knows that a combined machine is seldom a perfect one in any respect. Formerly, machines for reaping and mowing were combined; but farmers soon discovered that, although the first cost was a little more, separate machines were rather to be desired, since each afforded the advantages for its peculiar use without being burdened with more or less useless machinery, and was always ready for immediate use without changing the "rig." As already stated, a cow is but a machine for the production of beef, milk, or butter, and experience teaches that a cow of various qualities is never as profitable as one that is *bred, fed and cared for* with but one aim in view. Not every farmer is able to buy thoroughbred stock; but every one should have a definite object in view, or, as the years go by, he will gain nothing by way of improvement, and this is not progressive farming. Nearly a year is required for a cow to breed, and to allow this length of time to elapse without improvement in the stock is to lose forever a valuable opportunity.

In considering how the proper management of food may assist in the improvement of the stock of a farmer who has not the means to buy stock already improved, and how such an one may develope in his herd a greater proclivity for converting their food into beef, milk, or butter, as may be desired, let us take, for example, the "combined" cow, *i. e.*, one which excels in nothing, and see how she may be improved. First, for beef. The sire should be supplied with an abundance of rich and nutritious food, such as produces the most fat. He should be well cared for and furnished with comfortable quarters, and indulged in laziness, as that is conducive to flesh. When wanted for service he should be quite fat, that he may transmit to his progeny a tendency in that direction. The dam should be similarly fed, and should be taking on flesh, especially during the period of gestation, not so much for benefit to her, for she is already grown and no amount of feeding will benefit her. The object is to create in her unborn calf, with the assistance of the sire's influence, a greater tendency to fatten. This tendency thus transmitted is operative only during gestation, so that the opportunity for improvement in the young virtually ceases at its birth. For, at whatever stage of perfection the young of any animal may arrive, the result is due to the impetus received from the parents. Heifers should not be allowed to breed until at the age of three years, and very little account should be taken of the quantity and quality of milk. In fact many breeders allow the calves to run with their dams for three or four months and then dry them off. Mr. Price, a noted Hereford breeder, says: "Experience has taught me that no animals possessing form and other requisites giving them a great disposition to fatten, are calculated to give much milk, nor is it reasonable to suppose they should,—it is in direct opposition to the law of nature." By applying the above principles of feeding, we may reasonably expect to rapidly improve the tendency to fatten,—not that it will take less food to make a pound of beef, but because the growing generations have been better educated, if I may use the expression, to convert their food more entirely to the one purpose, *i. e.*, beef.

If it be desirable to improve on our "combined" cow, so that her progeny will be more profitable milk producers, the food for the sire should be plentiful, but of the kind which produces in the cow quantity of milk, rather than flesh, (*Miles on Stock Breeding, page* 90). Although bountifully fed, it should *never* be to the extent of developing fattening proclivities; nor should he be allowed to exercise to the developement of muscle. His general appearance should be feminine. The dam should be fed on such food as produces the greatest quantity of milk; and especially during the period of gestation, her feed must be bountiful. The milk secreations should be stimulated to their fullest extent, consistent with health, that the organs of lactation in the unborn calf may be more fully developed than those found in the normal condition of the dam. Heifers should breed early, —at least, when coming two years old—and trained to long milking periods, not allowing them to go dry more than one or two months before the birth of the second calf.

In breeding for butter, there is no particular difference in feeding from the method last described, except that the feed should be of a richer kind. The quality of the milk should be improved, regardless of the quantity. Some of the greatest milkers, giving forty-five quarts, or ninety pounds, per day, are, as butter makers, surpassed by one of the butter producing breeds which may give only twenty quarts per day. My own experience, which I think every dairyman and observing farmer will endorse, is that the cow which eats the most and keeps the poorest is the most profitable for milk or butter; while the cow which eats the least and takes on the most flesh is the most profitable for beef. Paradoxical as this may seem, it is in perfect harmony with the laws of nature.

Important as it is to understand the art of feeding, our best efforts may be defeated by an improper selection, which the condition of the farm and the demands of the nearest market, rather than any individual preference, should determine. Generally speaking, families of the bovine race may be divided according to their special advantages, as follows:

Those producing beef, Herefords and Short Horns.

Those producing milk, Holsteins and Ayreshires.
" " butter, Jerseys and Guernseys.

Farmers breeding for beef will tell you that such a calf gained three pounds of flesh per day for a year; but there is nothing to say about milk or butter. Those breeding milkers will show you a cow which gives from ninety to a hundred pounds of milk per day; but they have little, if anything, to say about the amount or quality of butter. The breeders of butter makers will show you a cow which produces three pounds of butter per day, but make no mention of the quantity of milk. Some one might suggest that it would be a good plan to couple the beefers with the deepest milkers, and again with the best butter makers, thus consolidating the desirable qualities of all breeds. The result would be merely a grade animal, probably little or no improvement upon our native cattle, and thus we should again have our "combined" machine. This plan of breading has often been tried but has as often failed. Each family has peculiar traits for which it is prized, and to combine these is, in a great measure, to destroy them. The laws governing the principles of breeding were ordained by the author of all creation, and took effect with the first breath of life. They are as changeless as the laws of force, gravitation, and light. How essential is it, therefore, that we should understand them and work in harmony with them? To attempt to alter or even modify them is to work discouragement and failure.

In representing the different breeds of cattle, it has not been for the purpose of elevating one breed above another. Each family, if selected for just what it is, will far surpass, in its peculiar capacity, all others. It is nothing short of absurdity to select milkers from a family which, for the last two or three hundred years, has been breed for beef, and *vice versa.* It would be equally consistent to choose the Clydesdale for the race course, or the Hambletonian for the dray. In my opinion, one of the most absurd theories advanced as a reason for selecting a large breed of cattle, with inherent beef proclivities, for the dairy, is "that when, we are through with them as milkers, we can realize more money on disposing of them for beef." Acting on this principle, a farmer, on being offered the choice of two cows, of 900 and 1,200 pounds weight, prefers the one heavier than the other. Now, it has been repeatedly proven by careful experiments, in England, Germany and America, that it requires of hay, in pounds, or their equivalent, two per cent. of the live weight, per day, to sustain life, without producing milk or flesh. At this rate, it takes six pounds of hay more per day to sustain the extra three hundred pounds of flesh; and the cost of this extra feed, at $15 per ton, would amount to $16.42 for the year. The average period of usefulness of a cow is ten years. So that the cost of sustaining the extra three hundred pounds until the cow had outlived her usefulness as a milker, would amount to $164.20; but, at that time, the three hundred pounds of beef would be worth only $12.00. Again, it is estimated that it requires three pounds of hay per day to produce one quart of milk. That is to say, after the animal has consumed two per cent. of her live weight, three pounds of hay, or its equivalent, in addition will be necessary to produce one quart of milk. Therefore, the six pounds of hay which is daily consumed to support the extra three hundred pounds of flesh, would produce in a cow weighing three hundred pounds less two quarts of milk, which, for three hundred days, the average milking season, would amount to six hundred quarts; and this, at three cents per quart would amount to $18 per year, or, for the ten years $180. As both cows originally cost the same, say $40, estimating the cost of keeping the smaller at $1 per week, and admitting that the larger cow gives ten quarts of milk per day, milking 300 days of the year, their accounts, at the end of ten years would be as follows:

	LARGE COW.		SMALL COW.	
	Dr.	Cr.	Dr.	Cr.
To cost,	$ 40.00		$ 40	
To cost keeping,	684.20		520	
	$724.20		$560	
By milk,		$900.00		$1,080
By beef,		48.00		36
		$948.00		$1,116
		724.20		560
Profit,		$223.80		$556

These figures serve to show the fallacy in supporting three hundred pounds of extra flesh for ten years for the sake of having that much more to dispose of at the end of that time. It also serves to show the greater comparative value of the cow which gives at a milking only one quart more than another. In ten years we go to a cow about six thousand times for milk; we can go to her but once for beef. It might be suggested that the calves of the larger cow would weigh at least twenty-five pounds each more than those of the smaller; but this, at four cents a pound, would amount to only one dollar per year; and, on the other hand, if the heifer calves were to be raised for dairy purposes, any one knowing the dams would, other things being equal, pay at least twice as much for the calves of the smaller cow as for those of the larger.

The reader may ask, what has the influence of food and the selection of a breed to do with soiling? Not so much perhaps as soiling has to do with such influence and selection,—by placing within the hands of the farmer the means by which he may best improve his farm stock, *i. e.*, an abundance of rich, succulent and nutritious foods during the entire year, which, when fed to stock possessing the tendency to convert it into the product demanded, will return to him handsome profits and afford the best and most economical means of enriching the soil for the growing crops. Thus he may feed his stock more bountifully, knowing that it will surely return to him in a well filled pail, or that it is being certainly stored away on the ribs of his fattening steers. He may feed with a more liberal hand, knowing that in the compost heap he is storing treasures, and that these will be committed to the fields, as to the hands of the faithful steward, which will return them again with usury. Once started in the right direction, forsaking all others, there will be no limit to the successful application of energy and skill. He may run with safety where, if his course were doubtful, to walk would be uncertain; and uncertainty robs a man of that vim and push which is so essential to success in any line of business. But, if on the right track, though he may stumble and even fall, he will yet recover. "Be sure you're *right*, then *go ahead*."

The Comparative Values of Grain and Forage Crops.

Says Lockhardt: "Good farming consists in taking large crops from the soil, while, at the same time, you leave it better than you found it."

Good crops make good manure, and good manure makes good crops. The value of grains and forage crops for animal food depends, principally, upon the amount of albumenoids, carbo-hydrates and fat which they contain; while their value for plant food (manure) depends chiefly on the amount of nitrogen, phosphoric acid and potash contained in them. Animals, in their consumption of food, retain little if any of those elements of it which constitute plant food; and plants consume little if any of those which constitute animal food. Thus, if a ton of feed should be plowed under for manure, it would be of no more value to the land than if it had first been fed to the stock, and none of its virtue had been allowed to waste while in the form of manure. Some plants and grains are very rich, or valuable, as animal food, while others are rich as plant food, and, again, others are valuable for both purposes.

Important as it is to know the influence of food, and the necessity of giving it to the proper animal to obtain the most profitable results, the highest degree of economy cannot be obtained without a knowledge of the comparative values of the feed consumed by the stock as animal and as plant food. The following tables will afford the farmer some curious and interesting facts, and some information which will assist him in making a most economical selection. The analysis, from which the values of the different foods are estimated, was taken from the late work of Dr. Emil Wolff, of the Royal Academy of Agriculture, Hohenheim, Wurtemberg. They represent the average result of numerous reliable analyses, and are sufficiently accurate for all practical purposes. The original analysis, representing the comparative proportions of different foods, is given in the number of parts found in 100 and 1000. From these I have estimated the number of parts, or pounds, found in one ton (2000 lbs.) Their values for animal food, in dollars and cents, is computed by estimating albumenoids at $4.00, carbo-hydrates at .80, and fat at $4.00 per hundred pounds. This is, doubtless, below their real value. Waldo Brown, of Ohio, estimates them, respectively, at $4.30, .90, and $4.35 per hundred. As the price of feed varies in different localities, these figures cannot be said to be absolutely and universally correct.; but they may serve to show the relative values of the different kinds of food. If, for instance, my estimate of timothy for feed, at $17.96 per ton is too low, all the others are proportionately too low. In calculating the values of the different grains and forage crops as plant food, I have taken the average of the prices given by several different authors, and find them to be about as follows: nitrogen, 20 cents per pound; phosphoric acid and potash, 4 cents per pound. I do not know where or in what form nitrogen can be had at 20 cents per pound, but that figure will suffice to give us the comparative values of the different farm crops as manure.

GRAINS.	Pounds of Animal Food per ton.			Value as feed per ton.	Pounds of Plant Food per ton.			Value as manure per ton.
	Alb.	Carb. Hyd.	Fat.		Nitrogen.	Phos. Acid.	Potas	
Field Beans,	510	910	40	$29.21	81.6	17.2	26.2	$18.04
Field Peas,	448	1046	50	27.38	71.6	17.2	19.6	15.78
Tares (Vetches),	550	844	54	30.91	88.0	20.0	16.2	19.04
Indian Corn,	200	1360	140	24.48	32.0	11.8	7.4	7.16
Wheat,	260	1352	30	21.51	41.6	15.8	10.6	9.36
Rye,	220	1384	40	21.41	35.2	16.8	11.2	8.16
Barley,	190	1332	50	20.25	32.0	15.4	9.0	7.37
Oats,	240	1218	120	24.14	39.4	12.4	8.8	8.72
Buckwheat,	180	1192	50	18.64	28.8	11.4	5.4	6.43

GRAIN VS. FORAGE CROPS.

GROUND FEED AND REFUSE.	Pounds of Animal Food per ton.			Value as feed per ton.	Pounds of Plant Food per ton.			Value as manure per ton.
	Albumen.	Carb. Hyd.	Fat		Nitrogen.	Phos. Acid.	Potas.	
Cotton Seed Meal,	660	352	324	$42.66	98.0	56.2	29.2	$23.00
Linseed Meal.	566	826	200	37.24	90.6	32.2	24.8	20.40
Corn Meal,	200	1360	140	24.48	32.0	11.8	7.4	7.16
Malt Sprouts,	460	894	50	27.55	73.6	36.0	41.2	17.80
Brewer's Grains,	98	222	32	6.97	15.6	8.2	1.0	3.48
Wheat Bran.	280	1000	76	22.24	44.8	54.6	28.6	12.28
Rye Bran,	200	1070	70	22.96	46.4	68.6	38.6	13.56
Rape Cake,	566	670	180	35.20	97.0	35.4	24.8	21.80

DRY FORAGE. (Hay and Straw.)	Pounds of Animal Food per ton.			Value as feed per ton.	Pounds of Plant Food per ton.			Value as manure per ton.
	Alb.	Carb. Hyd.	Fat.		Nitrogen.	Phos. Acid.	Potas	
Red Clover,	268	598	64	$18.06	39.4	11.2	36.6	$ 9.78
Timothy,	194	976	60	17.96	31.0	14.4	40.8	8.40
Lucern,	394	858	66	25.26	46.0	11.0	30.6	10.86
Tares, cut in blossom,	284	706	50	19.00	45.4	21.4	56.6	12.20
Peas, cut in blossom,	286	736	52	19.40	45.8	13.6	46.4	11.56
Orchard Grass,	232	814	54	17.95	31.0	8.2	26.4	7.58
Wheat Straw,	40	604	30	7.68	9.6	4.4	12.6	2.60
Rye Straw,	30	540	26	6.56	8.0	4.2	15.6	2.89
Barley Straw,	60	656	28	8.75	12.8	3.8	18.8	3.46
Oat Straw,	50	764	40	9.71	11.2	3.8	17.8	3.10
Pea Straw,	130	704	40	12.42	20.8	7.0	20.2	5.24
Bean Straw,	204	730	20	14.80	32.6	6.4	37.0	8.24
Cornstalks,	60	720	22	8.98	9.6	10.6	19.2	3.01

GREEN FODDER.	Pounds of Animal Food per ton.			Value as feed per ton.	Pounds of Plant Food per ton.			Value as manure per ton.
	Alb.	Carb. Hyd.	Fat.		Nitrogen.	Phos. Acid.	Potas.	
Grass,	60	258	16	$5.10	10.8	3.0	9.2	$2.61
Clover (red).	66	154	14	4.43	10.2	8.8	2.8	2.50
Lucern,	90	156	12	5.32	14.4	3.2	9.6	3.38
Tares (vetches),	62	152	12	4.22	11.2	4.6	12.2	2.90
Peas,	64	164	12	4.25	10.2	10.2	3.0	2.56
Oats,	46	176	10	4.14	7.4	3.4	15.0	2.21
Rye,	66	298	18	5.74	10.6	4.8	12.6	2.81
Corn.	22	218	10	3.02	3.8	2.6	8.6	1.20
Hungarian Millet.	118	300	30	8.32	20.0	2.5	17.0	4.78
Sorghum,	50	306	28	5.56	8.0	1.6	7.2	1.95
Cabbage,	30	126	8	2.52	..	..	..	...
Rape (leaves),	400	950	40	25.20	9.2	2.8	8.0	2.26

ROOTS, ETC.	Pounds of Animal Food per ton.			Value as feed per ton.	Pounds of Plant Food per ton.			Value as manure per ton.
	Alb.	Carb. Hyd.	Fat.		Nitrogen.	Phos. Acid.	Potas.	
Potatoes,	40	420	6	$5.20	6.8	3.8	11.4	$1.96
Turnips,	64	340	12	5.74	3.6	1.8	6.6	1.05
Field Beets,	22	182	2	2.31	..	..	..	...
Sugar Beets,	20	308	2	3.43	3.2	1.6	7.8	1.01
Carrots,	30	216	4	3.08	4.4	2.0	5.6	1.18
Pumpkins,	26	56	2	1.56	..	.	.	...

The compost heap is the farmer's bank, on which his drafts will be honored in proportion to the amount deposited therein. It is a mistaken notion that manure is manure, whatever it may be made of. The foregoing tables show a ton of clover hay to be worth $9.78 for manure, and a ton of corn meal only $7.16; but, as feed, the meal is worth about $6.42 more than the hay. The manure from a ton of wheat straw is worth about $2.60, while that from the same amount of cotton seed meal is worth $23.00. Clover hay is worth more than timothy, both as feed and as manure. The peculiar value of timothy for horses is that it contains a very large per cent. of carbo-hydrates (muscle forming food) and not so much fat. We also see that lucern and peas cut in blossom, make very valuable hay. There are very many interesting facts to be found in the tables which we have not the space to notice particularly. The relative value of each of the green forage crops is apparent at a glance. By a careful study of the tables, the farmer may make such selection of food as will be most economical and best adapted to the wants of his stock and his land. For instance, the price of corn meal is about the same as that of malt sprouts, but the latter is worth about $3.00 per ton more for feed and about $10.00 per ton more for manure. Last season I sold corn at the rate of $25.00 per ton, and bought cotton seed meal for $3.00, including freight from New York City; but the latter was worth nearly twice as much for feed, and three times as much for manure.

GREEN CROPS AS MANURE.

"Ordinary barnyard manure," says Mr. Harlan, in his work on Farming with Green Manure, "contains 10 lbs. of nitrogen, 5 lbs. of phosphoric acid, and 12½ lbs. of potash." By reference to the tables, we notice that a ton of green rye is worth just about as much. Millet and lucern show much larger values. Clover, also, is a valuable crop as a green manure. Its entire value is not shown in the analysis, as its roots, which are very numerous, are not taken into the account. The same is true of lucern. The advantage of ploughing under green manure is that it saves drawing and spreading. Rye, for this purpose, doubtless stands ahead of all other green crops, not on account of its superior value, but because it occupies the ground at a time when no other crop, except wheat, will grow. In ordinary land, there will be growing by the first of May, from 15 to 20 tons per acre. As a summer crop, cow peas are doubtless the most valuable. They are extensively grown for green manure on exhausted cotton and tobacco fields in the southern states.

LIQUID MANURE.

There is, perhaps, no branch in farm economy which receives so little attention as the saving of this most valuable fertilizer. Many farmers have brooks running through their yards, or have their yards on side hills or on gravelly soil. Thus stores of the most valuable plant food is lost.

The following table shows the number of pounds of nitrogen, phosphoric acid, and potash found in one ton of fresh dung or urine, and their comparative values:

	DUNG.				URINE.			
	Nitrogen.	Phos. Acid.	Potash	Value.	Nitrogen.	Phos. Acid.	Potash	Value.
Sheep,	11.0	6.2	3.0	$2.56	39.0	0.2	45.2	$9.61
Horse,	8.8	7.0	7.0	2.30	31.0		30.0	7.40
Cow,	5.8	3.4	2.0	1.87	11.6		9.8	2.71
Swine,	12.0	8.2	5.2	2.93	8.6	1.4	16.6	2.44

The analysis in this table is from Prof. Wolff, to which I have added the values, estimated according to the previous tables.

The two methods usually employed to save liquid manure are by absorbents, such as dry muck, sand, road dust, sod, shavings, saw-dust, forest leaves, cobs, etc., and by drainage from the stables into a cistern. The drainage from the yards may be pumped into portable tanks similar to street sprinklers, and thus distributed over the fields. Some farmers have hollow places in their barnyards, filled with absorbents, but they are usually great nuisances in open yards, and they should be under the stack or where the rain cannot fall directly into them. It is estimated by some that " the number of pounds of urine is at least double that of solid manure." I am inclined to think the proportion more nearly equal. In the case of horse manure, there is no attempt to save the liquid portion, and for every ton of dung, worth $2.30, saved, there is wasted $7.40 worth of urine. In saving this most valuable fertilizer, the soiling system affords the very best means, and I have no doubt that the saving of manure, in this direction alone, will more than pay the cost of the extra labor. The manner of saving by absorbents will be explained under the head of stables and their construction.

Liquid manure is used to quite an extent in England, where the soiling system has been adopted, as a top dressing, and enables the farmer to get from five to seven cuttings of feed from the same ground during the season. These dressings are put on immediately after each cutting. The urine is pumped from the reservoir into a tank or hogshead on a two-wheeled cart which is drawn by one horse. It is then distributed through a sprinkler or rubber hose attached to the tank. The liquid is never applied except after being greatly diluted with water. It is claimed that a man with cart and horse " will top-dress one acre per day, within a quarter of a mile from the barn." I doubt if the same force could draw out and top-dress, with solid manure, more than one quarter of that amount of ground per day. The German proprietor of eight acres, referred to by Morris in " Ten Acres Enough," who transformed the neglected and exhausted soil into a garden of immense productiveness and great profit, started with a capital of $3.00 and four pigs. The manure of his small stock, with the refuse of the family, was collected in a buried hogshead, there reduced to liquid manure and applied by means of a wheel-barrow. The results from this small beginning in the right way were so remarkable, that he soon added more stock, sinking a brick cistern in the barnyard into which the liquid manure from 6 cows and 2 horses was conducted, together with the wash from his pig-pen and yard. The manure heap, always under cover, was thoroughly saturated by means of a pump in the cistern, which was also used for filling a hogshead on wheels with the fertilizing liquid for application.

COMMERCIAL FERTILIZERS.

All known plants are composed of fifteen elements only. As, by the use of twenty-six letters of our alphabet, we are enabled to write them so as to express many thousand different words, so various combinations of these fifteen elements or letters in the alphabet of vegetation, produce many thousand different plants.

Most of our artificial manures are only special fertilizers, and supply the soil with only part of the plant food generally required. Let us suppose, for example, that we wish a certain field to produce a crop of wheat, and that, in order to grow the crop will require 5 of the 15 elements or letters in our vegetable alphabet. Let us represent these letters as W H E A T. If but one of the letters is missing, the word is incomplete, and the soil fails to spell the word we require; in other words, the crop is a failure. How is a farmer to know what one of the letters is missing? He must seek his answer in the ground by analyzing it. Can the farmer analyze each kind of seed and the soil in which he is to sow it to determine their fitness for each other? And must he repeat the experiment with other seeds for succeeding crops, or with other soils for the same crop? You say that this is impossible. Then is it also impossible for him to apply special fertilizers intelligently or profitably; for, while such may do well in one field for a certain crop, and show a good return for the money expended, the same in an adjoining field may amount to nothing.

The application of complete fertilizers is, at least, a step in advance, because, if the soil is supplied with all the elements necessary to produce a crop, one is more certain that the missing letter is supplied. We will say, in the case of wheat, that all the letters, but T, are present, and that the missing letter represents potash, worth four cents a pound. The soil being supplied with the other four elements, a farmer will pay $40.00 to $60.00 per ton for a complete fertilizer whose potash is worth $4.00 or $5.00. Thus it often occurs that the application of a little lime, salt, ashes or plaster, side by side with fertilizers costing $60.00 per ton, produces equally good results, and not because, as some farmers suppose, the fertilizer was worthless, but because the soil was already possessed of all the elements afforded by the fertilizer, except that simple one which a much cheaper substance could as well afford. It is the testimony of many farmers that the use of fertilizers sometimes weakens, rather than strengthens the soil, by stimulating the available stores of plant food to an unusual degree, and producing an effect similar to that of stimulating drink upon a person who is, by its influence, able to perform an unusual amount of work, but, when the influence is withdrawn, is in a weaker condition than when he began. Many of these fertilizers are, in my opinion, too expensive, considering their values as manures. I find, in applying the same rates for valuation as were used to compute the values of the different crops for manures, in the preceeding tables, that a brand selling for $40.00 per ton is worth but $12.27; another selling for $35.00 is worth only $9.00; and one selling for $60.00 is worth but $33.40. If these estimates are too low, then the values set to the different grain and forage crops, also, are too low. If analysis is worth anything, a ton of wheat bran, fed and made into manure, is worth ($12.28) as much, ton for ton, as the fertilizer above referred to as costing $40.00; and a ton of clover hay (worth $9.78) is as good as the brand selling for $35.00; a ton and a half of cotton seed meal is worth after feeding, $34.00.—about the same as the brand selling for $60.00 per ton. I must not neglect to say that there are some brands which do much above their full and actual values as manures; but, even in those cases, the same amount produced by feeding would be far more economical, because thus the farmer receives a double profit,—first, by the production of beef, milk, or butter, and again, by the production of manure. I do not wish to be understood as utterly condemning commercial fertilizers, but I do mean that the proper application of barn-yard manure is the surest and most economical way of enriching our soil. Says Prof. W. O. Atwater, " Stable manure contains all the ingredients of plant food. It is a complete fertilizer. Nor is this all. It improves the texture of the soil, it tends to regulate the supply of moisture, and it helps to set free the stores of inherent plant food which every soil contains." As a means of attaining large quantities of manure at "a cheap and easy rate," the soiling system affords the greatest advantages.

STRAWS, WHICH SHOW WHICH WAY THE WIND IS BLOWING.

England has been obliged to adopt the soiling system in order to increase the number of farm stock per acre,—in other words, that she may grow larger crops of grain. In France and Germany, soiling is the rule, pasturing the exception; and the number of their live stock has been greatly increased since the introduction of the sugar beet industry. It is hardly necessary to add that the productiveness of their soil has increased correspondingly. The people of these countries have no uncle rich enough to give them each a farm, and, therefore, to supply the demands of their increasing population, they have been obliged to increase the yield of their present possessions by doubling and tripling the productiveness of their soil. As in a crowded city they add to the capacity of their houses and factories by building up story above story, so the farmers of these older countries have been obliged to *build up their soil*, until they have farms two, three, or four stories high; that is to say, they have increased the productiveness of the soil until one acre is made to produce what formerly required two, three, or four acres. There is, I venture to say, scarcely a farmer east of the Mississippi who would not be glad to know how this is done. The secret is an open one, *i. e*; by growing enormous crops of roots, clover and artificial grasses, and by keeping the greatest possible number of farm stock to consume them, thus making large quantities of manure which, in their turn, produce large crops of grain. In fact, by adopting a regular system of soiling, the majority of English farmers rent their farms and pay, says Hon. H. F. French, "an average of $10.00 per acre (yearly) for the entire farm, on land which has been under the plow for centuries." That he is able to pay this rent, and $10.00 per acre for manure, and $10.00 more for expense of cultivation, support himself and family comfortably and add something to his income, shows what the system has done towards attaining and maintaining a high state of fertility in the soil. How many farmers in this country, with one, two, or three hundred acres of land, could, under the present system of growing crops and feeding stock, pay a rent of $10.00 per acre and live? I will not attempt to answer. I do believe, however, that, if we were obliged to do so, *we could*, and perhaps even more easily than our English neighbors, for our land is certainly naturally as productive as theirs. The following figures show that the advantage is in the fertility of the soil. The United States' census report for 1850 shows that the average yield of wheat in this country is 9⅛ bushels per acre, while in England it is 28, and in Scotland 29½. There is another item in the comparison with the English farmer which I cannot pass without mention, *i. e.*, his whole time, attention and capital are devoted to his farm; he has no other source of income. A large number of American farmers, I am ashamed to say, devote the greater part of their time, but only half of their attention and little or none of their capital, to their business. I refer now to farmers with capital, with a few hundred dollars, which they have saved out of their farms in prosperous times, but, instead if keeping it employed in their business as they should, have invested it in uncertain stocks, promises to pay, etc.

Money, *as a concentrated force*, is the lever which moves the commercial world. It sends the railroad puffing and blowing into every hamlet which beckons it, over a Niagara, through a mountain tunnel, under an English Channel,—anywhere. It cobwebs the land with electric wires, until New York,

Boston and Philadelphia are our next door neighbors, while England, France and Germany are just across the way. But money, as a divided force, falters before a streamlet, stumbles at a molehill, and comes to a dead halt at a stagnant pool. To have a business out of which the capital is taken to run another man's trade, is to rob that business of its lever power. A seductive speculation is a "Will o' the Wisp" to the farmer who follows it. Thus his attention is divided between cultivating the roots of evil and those of his growing crops.

But to return to our subject. We were considering the condition of the older countries and the manner of improving their soil. We need scarcely reflect to discover that our own country is fast approaching the same conditions. Land is increasing in value. "Out west" is no longer out west, and farmers are beginning to realize, as never before, that the productiveness of our soil must be increased, or we "be driven to the wall." Let us, then, learn the lesson indicated by the straws which point to a system of farming by which the productiveness of our soil will be increased instead of diminished. This means to increase the number of farm stock, which means to have more manure, which means an increase of power in the soil, which means ability to grow more profitable crops, which means a larger income, which means better books, better education, warmer houses, a better seat in the cars, at the lecture, and in the church, which means *independence*. All these good things are the results of better farming, which is itself the result of

SOILING.

What induced me to adopt the system? When I was a lad, the farm on which I now reside was known as one of the best in the county. The deep gravelly loam of the valley never gave so bountiful harvests of spring crops, and the side hills facing the rising sun, yielded, of wheat, 35 to 40 bushels per acre. The pastures afforded the most nutritious grasses for flock and herd. On taking possession of the place years later, I was surprised at the change. The number of cattle had been decreased by half and the flock of sheep had disappeared entirely. My first wheat crop from the field once considered the best, and over which, as a boy, I had driven the reaper, cutting 40 bushels per acre, measured but 15 bushels per acre. I acknowledge a sad disappointment at the time, and I made the following calculations, from which the difference will be more apparent.

Statement showing the Cost, and the Profit and Loss of growing 15 and 40 Bushels of Wheat per acre.

	15 BUSHELS.		40 BUSHELS.	
	Dr.	Cr.	Dr.	Cr.
To Fitting the ground,	$5.00		$ 5.00	
To 2 bushels seed, at $1.10 per bu.,	2.20		2.20	
To Interest, at 7 per cent. on 1 acre,	8.75		8.75	
To harvesting and drawing to barn,	1.75		2.00	
To threshing, &c., at 6 cts. per bush.	.90		2.40	
To marketing, 1½ cts. per bush.,	.22		.60	
By cash for wheat at $1.10 per bush.		$16.50		$44.00
Total, - -	$18.82	$16.50	$20.95	$44.00
Balance, - -	16.50			20.95
Loss per acre,		$2.32		
Gain per acre,				$23.05

For the 16 acres, the total loss on a product of only 15 bushels would be $27.12, while with a product of 40 bushels there would be a profit of 368.80. The difference, per acre, in cost was only $2.13, but the difference in income was $27.50. What had become of the farm? The land was all here, but the farm,—where was it? I soon discovered. It had gone to New York and Boston,—been sold by the bushel. The railroad was little by little, stealing it away ; the canal was peddling it out along the wharves of the metropolis. Nor was this all. The food had been taken out of the mouths of the stock, their number yearly growing less, and the few remaining, not receiving a proper supply of food, failed to make a profitable return. In fact, the whole superstructure of the farm was undermined. Thus it is that many broad acres, once productive, are to-day as deceptive as the apples of Sodom. In taking a general survey of the subject, I came to the conclusion that the only way of redeeming the fertility of my soil would be by the proper application of barnyard manure. In conversing with the most prosperous farmers, I found invariably that their success was owing to their faith in manure. There was already more stock on the place than could be properly fed, but not enough to keep the farm even in the condition in which I found it. And this, I said to myself, they call farming,—"the most independent life that a man can lead!" It seemed to me then the most dependent ; and it all depended on a single question,—if the soil is productive you may depend upon a good crop ; if not, you may depend upon a poor one. A ten-acre clearing, full of stumps, which will raise wheat at a profit, affords more independence than a hundred-acre farm, where it costs more to produce a crop than it does to bring it to market. About this time I chanced to read a work on "Soiling," by the Hon. Josiah Quincy, in which he stated how the farmer, by adopting the system, would be enabled to keep four head of stock where, by pasturing, he could keep but one. This seemed to be the key by which I might unlock the storehouse of the farm's capabilities, and thus I might again see these fields yield profitable harvests.

ADVANTAGES.

By soiling is meant a system of feeding farm stock with grass, or other green forage, cut and brought to them from the field. The advantages which it offers to the farmer are numerous, but the principal reasons why it should be adopted may be enumerated, as follows:

1st.—Saving of land.
2d.—Saving of fences.
3d.—Saving of food.
4th.—The better condition and comfort of farm stock.
5th.—The greater production of milk, beef, or butter.
6th.—The increased quantity and quality of manure.
7th.—The increased productiveness of the soil.

The disadvantages of the system as compared with pasturing are, as follows:

1st.—It requires extra labor.
2d.—It keeps the stock in close confinement.

That we may satisfy ourselves of the truth or falsity of the above assertions, *pro* and *con*, let us consider them severally in their order.

SAVING OF LABOR.

Says the Hon. Josiah Quincy, whose experience in soiling covered a period of eighteen years, "One acre soiled from will produce at least as much as three acres pastured in the usual way, and there is no proposition in nature more true than that any good farmer may maintain, upon 30 acres of good arable land, 20 head of cattle the year round;" he adds, "My own experience has always been less than this,—never having exceeded 17 acres for 20 head. * * * I have kept the same amount of stock, by soiling, on 17 acres, that previously required 50." Says Mr. H. Stewart, of New Jersey, in *The Country Gentleman*, " By soiling, J. D. Powell, of Winchester Co., keeps 100 cows on 100 acres," and adds, " with complete soiling, I have myself kept 14 cows on 11 acres the year round, with the help of a few loads of brewers' grains and some bran and meal." Where land is in a high state of cultivation, some farmers claim to keep as many as seven or eight head by soiling, where they could keep but one by pasturing. I think, as a rule, it is safe to say that land in an ordinary state of cultivation will support four head by soiling to every one pastured. My own experience in soiling, most of the time, is that, for every head formerly pastured, I was able to keep three by soiling, and, at the same time, for every acre formerly plowed to crops, I was enabled to plow two. My farm contains 100 acres of arable land inside the fences. Previous to adopting the soiling system, it was our custom to cultivate, on the average, forty acres, leaving sixty acres for pasture and hay. The average number of live stock (1,000 lbs. each) that was supported from the sixty acres, including the coarse fodder from the cultivated portion, was twelve. By soiling, during the last two seasons, I have been able to keep, on the average, the equivalent of 36 head weighing 1,000 pounds each, as follows; 13 cows, 5 yearlings, 4 calves, 4 horses, 2 colts, and 70 long-wooled sheep, —a total of 98 head. During the same time, we have had under cultivation (exclusive of land devoted to soiling crops) 70 acres. The remaining 30 acres, therefore, supported yearly the equivalent of 36 head weighing 1,000 pounds each. Some of the coarse fodder,—barley straw and stalks, was consumed during the winter, and, in addition, I have bought a few tons of cotton seed meal and bran to keep up a good flow of milk during the winter. I feed no grain while soiling during the summer. Thus, where, by pasturing, it required 60 acres to support 12 head, by soiling, I was able to keep, in better condition, three times the number on just half the amount of ground. It will also be seen that, where, by pasturing, I could cultivate but 40 acres a year for crops, by soiling I harvested 70 acres, nearly doubling my acreage *without buying it*. It may be asked, how could a farm maintain such a heavy cropping? I would reply that by soiling, we were producing and saving three times the amount of manure, the value of which certainly was double that made while pasturing, and thus, while we were cultivating nearly twice as many acres as formerly, there was produced nearly six times as much manure.

Where the system is practiced, nearly or quite all of the inside fences may be removed, and the land before occupied by them may thus be devoted to crops. About May 1st, 1880, we turned twelve head of milch cows to pasture in a field containing 4½ acres. At the end of the fourth week, I was obliged to take them out as they were shrinking greatly in flow of milk and the pasture exhausted. They were turned into the sheep pasture until June 7th, when we began soiling, keeping the same 12 head upon 4 acres during the next four months, making 1 acre soiled from, fully equal to 4 acres pastured from. From the 4½ acres pastured, I obtained, at 50c. per week for each head, a feeding value of $24.00,—from the 4 acres soiled, at the same rate, a feeding value of $96.00,—a difference in favor of soiling over pasturing, on 4 acres, of $72.00, leaving the increased value of manure to pay the expense of growing and feeding the crops. This establishes no rule, but is sufficient to illustrate the great saving of land and the economy of feed.

SAVING OF FENCES.

In some sections of the old countries where the soiling system is generally practiced, the farmers have done away with interior and boundary fences, setting land marks to indicate lines, and thereby working every foot of land. Says Mr. A. W. Cheever, in *The Country Gentleman,* "Another great advantage I find in soiling over pasturing is the saving of fences. None of my mowing or cultivated fields are pastured at all, so I have been enabled to dispense with all inside fences, and lately have been giving up the use of road fencing also."

No farmer will disagree with me in saying that farm fences are great nuisances, harbors for rats, mice, and vermin, most convenient places for noxious weeds and grasses, and great hindrances in every stage of farm work. For instance, if we wish to cultivate two fields adjoining each other but separated by a fence, we must stop and turn about as we approach the fence from either side in in plowing, harrowing, cultivating, rolling, drilling, reaping and raking. Thus, in growing a crop of corn, with a fence forty rods long, it would require about 1,500 or 1,600 turnings, and for wheat 1,200 or 1,400, according to the mode of culture. All this wastes time, besides trampling down the ground and crops. As Mr. Quincy says, " The whole farm may be divided and cultivated with precise reference to the state of the soil, when the plow runs the length of the furrow determined by the judgment of the proprietor." His farm at one time had 5 miles of interior fence, (equal to 1,600 rods), of which he says, " I have now not one rod of interior fence; of course, the saving is *great, distinct and undeniable.*" My own farm was at one time divided into 17 fields, which required over 1,000 rods of interior fence, the interest on the cost of which would pay the taxes on the entire property, or pay for all extra labor of soiling 12 or 14 head of stock, to say nothing of the cost of yearly repairs. I built some 300 rods of fence soon after coming on the farm. It hardly made a showing compared with what was needed. It would have required an outlay of at least $1,000 to put all the fences in proper shape, and for what? Simply to keep 12 head of stock from destroying the crops. Each field must be fenced, for, by the rotation of crops, each field was in turn pastured.

Reader, if you are a farmer, don't build another rod of fence until you have given the soiling system a fair trial and find it a failure. My farm has now but 7 fields, and I am yearly reducing the number. Says D. S. Curtis, on the cost of fencing, in the *Agricultural Report* of 1859, "The most ordinary plain board fences cost from 8 to 10 shillings per rod, and more in many places, while rail fences are often still more costly. But, taking the lowest estimate, $1,00 per rod, the expense of enclosing an eighty-acre lot would be $480.00; two cross fences, one each way, throwing the lot into four 20-acre fields would cost $240.00 more, a larger sum than the value of the land, in many localities." As Mr. E. W. Stewart says, "Soiling effectually settles the fence question."

THE SAVING OF FOOD.

There are several ways in which farm stock destroy their feed while at pasture,—by tramping it under foot, by their dung and urine, and by lying on it. The more productive the pasture, the greater the loss. Just how much is wasted by these means, I do not know. Some estimate it at one-third, others at a half. Another item of more or less importance is that it is not so exhaustive of the soil to grow a crop of hay from it as to use it as a pasture, especially if the grass of the pasture be closely cropped, thus having the soil more exposed to the sun. All these objections are overcome by soiling. The food may be cut at just the proper time, when the leaves and blossoms have reached their full development. It is often noticed that, here and there in a field, patches of distasteful grasses or noxious weeds are left untouched by the stock, except in case of great hunger, and allowed to ripen and go to seed. The seed is scattered about the field and pressed into the soil under the hoofs of the feeding stock. In time the pasture thus becomes only a garden of weeds. This would never occur were the practice of cutting adopted. Mr. Youatt, an English author, says, in his valuable work, The Complete Grazier, "If a close consumption of plants is the object principally to be regarded, it is evident that the benefit to be derived from soiling will be very great; for experience has clearly proved that cattle will eat many plants with avidity, *if cut and given to them in the barn*, which they would never touch while growing in the field."

What may be left by the horses when soiled, is eaten with relish by the cows ; and, if any is left in the mangers of the cows, it may be given to the pigs ; and thus, by soiling, not a particle of feed need be wasted.

The better Condition and greater Comfort of Farm Stock.

The difference in the condition of animals soiled and those pastured shows a decided advantage in favor of soiling. During the heat of the day, they may be kept in cool stables, darkened to exclude the flies which, during the greater portion of the season, torment them and drive them into a state little short of frenzy. Here they are also protected and sheltered from driving storms and the burning sun, and are secure from jumping into fields of growing grain or fruit orchards, and from injuring themselves or causing their own death by over-eating. They are also protected against eating noxious weeds, which often injures the quality of milk and butter, against drinking muddy and impure water, against worry and annoyance from dogs, and, above all, against *hunger and thirst*. These, and all other evils incident to pasturing, are reomved by proper management and well constructed stables and sheds, and reveal the strong points in favor of the soiling system.

Mr. E. W. Stewart, in relating an experiment to satisfy himself in regard to the comparative conditions of cattle soiled and pastured, says, in substance, that he put five steers and heifers into a good pasture for three months during the best part of the pasturing season, while others of the same age and condition were soiled, and that on comparison, at the end of the three months, those soiled were found to be in decidedly the better condition. The same cows, pastured one season and soiled the next, proved that their condition was better when soiled. His cows, soiled for five successive years, kept in good condition and uniform health.

It must be observed that all varieties of ruminating animals are naturally averse to any great amount of exercise to obtain their food, but, if it be supplied with abundance, fill themselves quickly and lie down to enjoy chewing their cud, and it is then, properly, that the animal is feeding.

On this subject, I cannot do better than to quote from H. A Willard, A. M., in his valuable work on "Dairy Husbandry" (p. 48): "It is not necessary that cows should be continually feeding, for we can see from the peculiar structure of their stomachs that nature intended a considerable portion of time to be spent at rest, that the process of rumination and digestion be perfected. The first stomach seems to be simply a receptacle for storing up a quantity of food to be used and enjoyed at leisure. The food, as it goes into the first stomach is very imperfectly masticated. After having filled this receptacle, the animal rests from her labors and is now prepared to enjoy her food, which is thrown back, in small quantities, into the mouth, where it is chewed and then goes into the third and fourth stomachs to be properly assimilated and digested. Hence, rest is required, and to deprive the animal of a comfortable resting place, or

to drive her out in the hot sun while in the act of rumination or masticating her food, is not only cruel, but a piece of intolerable stupidity. * *
The principle is true, whether acknowledged or not, that the more comfortable we make our milk stock, the better will be the results. If, during the heat of the day, cattle seek shade and lie down to rest, their quietness, comfort and enjoyment will add more to the milk pail than food taken in discomfort and excessive exercise."

It is also essential that the stock should be supplied with water, pure, plentiful and near at hand; for, if obliged to travel some distance to get it, they will "go dry," or wait until severe thirst compels them to seek it, which is not only a source of annoyance to the stock, but, in the case of dairy cows, a loss to the owner. Milk cannot be made without water, and when it is secreted largely, a large amount of water is absolutely required. Milk contains at least 75 to 80 per cent. of water. Here again the soiling system shows its superiority, since it affords the easiest and most economical way of supplying stock with water. Instead of having to furnish it in every field where rotation of crops and pasture is practiced, one good never-failing well or cistern at the barn is all that is required. On many farms, this is indeed a most valuable consideration. It has been my practice during the past three years to stable my cows during the day, letting them have the run of the barnyard during the night to enjoy the cool air and exercise as much as they please, and take generous drafts of cool water from a large tub supplied by a living spring. The yard is kept well littered with straw. The stables are in the basement of a barn 30 by 40 feet. The windows are darkened during the season when flies are troublesome. The cows are provided with comfortable bedding of shavings, or straw, or both. They seem to take solid comfort eating and chewing their cud. Observe those in the field wearily seeking their food, parched by the oppressively hot and dry air, and fighting flies until nearly wild; and then step into a cool, dark stable from which the flies have been excluded, and you will see a picture of comfort that I could not describe were I to devote a whole chapter to the attempt. Some are quietly feeding, some fast asleep, while others are diligently chewing their cuds, rolling them about in their mouths like some delicious morsels exuding nectar. In the expressions of their faces and in their clear bright eyes, you may read the unmistakable signs of *contentment, comfort* and *health*.

If you wish to thoroughly test this question, turn half of your cows to pasture and soil the rest. Any time during July, August, September or October, notice the cows coming from the field at milking time. They look tired and hungry, as if coming from a hard day's work, guant and thirsty, with languid step and melancholy look. Now open the stable door and let out the the cows that are soiled. They act more like "school boys from their books," each head erect, step sprightly, hair sleek, stomach full, and ready for a frolic. This is no fancy sketch,—indeed, I feel as if I had failed to fully represent the great contrast as I have witnessed it in my own yard. I feel safe in saying that I think no candid farmer, however predjudiced he may be against stabling his cows in summer, would need further proof to convince him that, so far as the healthful condition and greater comfort of the stock are concerned, the soiling system affords the most gratifying results, and adds materially to the

GREATER PRODUCTION OF BEEF, MILK AND BUTTER.

On this question, there can be but one opinion, *i. e.*, that to produce either beef, milk or butter, the result will depend upon the amount of food consumed, and the profit will largely depend upon furnishing our stock with an *abundance of succulent food during the entire year*. To accomplish this *independently of parched pastures, and drouth*, is *not* a difficult matter by the practice of soiling.

The following testimony as to the superiority of the system, was given by Mr. E. W. Stewart, in an article in the *Country Gentleman*, (1877, p. 42.) :

"We shall find the same reasons apply, in still greater force, in the growth of beef or mutton. Animals intended for slaughter should have different treatment from those whose value depends upon the development of muscle. Those reared for labor need much exercise, as well as appropriate food, for strengthening the bony and muscular system; but those intended for human food, need only so much exercise as promotes health and a vigorous appetite. And as we have seen soiling give a greater command over the supply of food at all times, so when properly conducted it must afford a greater certainty of rapid growth. We have easily grown calves, on green food fed in the yard, together with skimmed milk, that weighed 700 pounds at 10 months old. We have uniformly found this system more favorable to the growth of young animals than pasturing—that less grain or milk in addition is required to produce equal growth. And steers and heifers, during the second year, will make a steady and uniform growth on the full soiling system, with the liberty of a small lot for exercise. Animals for beef or milk are not grown for muscular strength, and require only a moderate amount of exercise. They need most full feeding, fresh air and kind attention. The skillful feeder has here an opportunity to observe the wants of each animal, and may always supply them. * * *

There must be no standing still if a steer is to gain two pounds for every day of its age up to 900 days. German and French beef growers adopt largely a strict soiling system, and produce a higher average weight at a given age than any pasturing people has attained.

Soiling also offers the opportunity of doing the principal fattening in warm weather, when not more than 75 per cent. of the food is required to make the same gain as in winter. We tested the the comparative effect of soiling and pasturing on the same class of animals, by putting five two-year-old steers and heifers, weighing 4,500 pounds, into a good pasture, while five, of the same age and condition, weighing 4,450 pounds, were soiled, with exercise in a small yard, and at the end of four months, those in pasture had gained 625 pounds, and the five soiled had gained 750 pounds, with nothing but green soiling food, making the two lots equal in kind of food. The pasture although good and abundant when the experiment began, did not continue throughout equally good on account of dry weather, while the soiling food was given in equal abundance to the end."

Mr. Brown, of Mankle, Scotland, tried the comparative merits of soiling and pasturing in fattening 48 steers, equally divided. The 24 soiled brought £377, and the 24 pastured £342,—a difference in favor of soiling of £35, or a profit of over $7.00 per head, to say nothing of the saving of land and the increase of manure.

In regard to the greater production of milk Mr. Stewart relates "the most remarkable test of the two systems, published by Dr. Rhode, of the Eidena Royal Academy of Agriculture, of Prussia. It was conducted through seven years of pasturing and then seven years of soiling. Mr. Hermann is the experimenter. The pasturing began in 1853, and ended in 1859—the soiling began in 1860, and ended in 1866. From 40 to 70 cows were pastured each year, and a separate account kept with each cow. The lowest average per cow is 1385 quarts in 1855, when seventy cows were kept, and the highest 1941 quarts in 1859, when forty cows were pastured, and the greatest quantity given by one cow was 2988 quarts. The average increased during the last four years from 1400 to 1941 quarts. The average per cow for the whole seven years of pasturing was 1583 quarts. In the soiling experiment twenty-nine to thirty-eight cows were kept, and the lowest average per cow was 2930 quarts in 1862, and the highest per cow 4000 quarts in 1866. The highest quantity given by one cow was 5110 quarts in 1866. The average per cow for the whole seven years of soiling was 3442 quarts. The yield of the same cow is compared for different years. Cow No. 4 gave in 1860, 3636 quarts; in 1863, 4570 quarts: in 1866, 4960 quarts. Cow No. 24 gave in 1860, 3293 quarts; in 1863, 4483; in 1866, 4800 quarts.

Many of these cows were the same in both experiments; and it will be seen that the same cow increased from year to year, showing what full feeding will do, and also another important fact, that this full feeding was conducive to the health of the cow during the seven years."

Dr. Wright says of soiled cows that they "will at least equal, if not surpass those kept in the usual way, in both quantity and quality of milk, and the dairyman, by adopting this method, finds his profits enhanced nearly one-fourth." An English author says, "The cows used to stall feeding will yield a much greater quantity of milk and will increase faster in weight when fattening than those which go into the field."

I have made repeated experiments which satisfy myself in regard to the increase of milk and butter, and, with the exception of the first month or two (May and June) I have never failed to get better results from the soiling system. There is, doubtless, no system of feeding (with forage alone,) that will excel in the production of milk and butter, that of a bountiful pasture of nutritious grasses during the months of May and June. But, from this time on, the soiling system has a decided advantage since, as soon as the pasture begins to fail, there is a corresponding failure in the flow of milk.

The author of "Ogden Farm Papers," in the *American Agriculturist*, has a very interesting article on the subject of soiling, in which he says, "The product of the cows will be more in the case of soiling than in the other. In June, I was making a very satisfactory amount of butter. So were the pasture men all around me. Now that the drouth has (in spite of passing rains) begun to affect the pastures, their product is falling off and by September, will be materially lessened. My product is increasing week by week, until, from the same number of cows, it is now over ten per cent. more than in June, and, an experience of previous years has shown, it will be fully ten per cent. more in September than it is now."

INCREASE IN QUANTITY AND QUALITY OF MANURE.

No farmer needs to be told that, if he has an abundant supply of manure, he can raise large crops. The want of it, more than any other one thing connected with farming, makes thousands of farmers and their families slaves to unremitting toil, drudging through life, selling away annually the fertility of their soil,—their birth-right, when, if one quarter of the labor that is spent in trying to subsist by cultivating exhausted soils, were turned to accumulating a restorative, independence would take the place of dependence, and the farmer enjoy all the comforts implied by well-filled barns and granaries, and, instead of dropping into a premature grave, live to enjoy a green old age, with energies and faculties unimpaired.

Manure is the very life and soul of husbandry. It is the basis of vegetable production,—the substructure on which alone the farmer can hope to build successfully. The attainment of manure by the soiling system is one of the greatest and most characteristic benefits to be derived from its practice, and the amount which thus naturally accumulates far exceeds all anticipation. All who have had practical experience agree, so far as I have been able to learn, that the value of the manure made under this system, when properly conduct-

ADVANTAGES.

ed, is worth, at the very least estimate, twice as much as that made while pasturing where it destroys as much feed as its virtue enriches the soil. A great part is lost by falling upon rocks, among bushes and in water courses. It is evaporated by the sun. It is washed away by the rains. Insects destroy a part. The residue (a dry hard cake) lies sometimes a year upon the ground, often impeding vegetation and never enriching the earth in anything like the proportion it would do if it had been deposited under cover. My own experience in keeping 12 cows, allowing them the run of the barnyard at night, is that they produced, while in the stable, one ton of manure every three days. Sufficient shavings are used to absorb the liquid portion. We will say that what was made in the yard during the night was no better than it would have been if dropped in the field, but that made in the stable and kept under cover was worth twice as much. The one ton would be worth at least $2.00, but, if made in pasture, only $1.00. The increased value of stable manure would therefore be 33 cents per day. I have found that it requires 2½ hours extra labor per day to soil 12 head of milch cows, and during the last three seasons this labor has cost me 6 cents per hour, or 15 cents per day; which affords a profit in the manure alone of 18 cents per day. This profit is sufficient to pay all other expenses. Therefore, I do not hesitate to say that the increase in quantity and quality of manure is ample to pay the expense of all extra labor incurred by soiling over pasturing.

But the saving of land, of fences, of food, the better condition and greater comfort of farm stock, the increase in the production of beef, milk and butter, and the attainment of manure, are all subservient and subordinate to the one prime object and benefit to be derived from the system, *i. e.* ;

THE INCREASED PRODUCTIVENESS OF THE SOIL.

I think no *farmer* could, within the last few years, travel through the southern states and see the deserted, fenceless farms, once the abodes of farmers whose soil and wealth were unparalelled,—but now the haunts of the emancipated slaves,—without recognizing at once that it is not on account of the political condition of affairs (as the politician views it), not so much on account of the ignorance of the black (as the learned men inform us), but for *the want of fertility in their soil.* Where is there a fertile country in any civilized land, which is not prosperous and peaceful and desirable to dwell in? There is probably no section of land in the world that nature has made more desirable to live in and has better fitted for the highest development of agriculture, than these southern states,—a climate unequaled, navigable rivers, in number and length unsurpassed, open to the traffic of the world,—in fact, a better site for the location of a Garden of Eden could scarcely be found. The unfortunate sons of sires whose curse it was to have the means to destroy without the knowledge to save the soil, decline to accept these exhausted fields, and push on to the west, leaving the negro, without the means to improve or the knowledge to restore it, to lounge about the premises in shiftless and melancholy despondency.

There are many sections in the north where the condition of farms is but little better. Sons are refusing to follow their father's vocation. Why? Because there is no money in it. The fields have been robbed of their fertility,—sold by the bushel in the markets of a selfish and greedy world, and washed into the sea. It is no wonder that farmers' sons decline to become slaves to the drudgery that is required to obtain a living from their worn out, exhausted soils. The land is therefore left to the emmigrant who, coming from a land of want, may be willing to live half fed, half clothed, without educating his children, and whose wife and daughters constitute the "help" of the house and farm. Such may manage to live, and, if hard physical labor will do it, manage to pull through. But an American, educated at the present day in our common schools, *cannot* and *will not* follow farming *unless* he can see money in it. He must have books, papers, recreation, and distinction. He is willing to toil night and day to secure these, but, if it is not to be found in the soil, he will seek it elsewhere.

The first, greatest, and most important question that can occupy the attention of eastern farmers, is, in my opinion, how to restore the fertility of our soils; and, as to the western farmer, how he may preserve it. If the reasons I have already given have nothing in them of sufficient importance to induce the farmer to adopt the soiling system, the fact that it affords *the surest and most economical way of increasing the fertility of his soil*, should lead him to give the system a fair and thorough trial. And, again, to the farmer who wishes to add more acres to those he already owns, the soiling system affords a certain means of doing so without buying more land. In my own experience, as already shown, soiling has doubled the acreage of my cultivated land, it has increased the quantity of manure three times, and the quality of it five or six times. I find, in looking about, that we have about as many head of stock as are generally found on a farm of 300 acres, and that we cultivate as many acres for crops as are cultivated on the average farm of 200 acres. My farm is by no means in a high state of cultivation, but perhaps a little higher than the the average. The system has done no more for me than it may do for any other farmer who will conform to its requirements, which are simple but exacting.

OBJECTIONS TO SOILING.

EXTRA LABOR.

The extra labor of soiling over pasturing is greatly magnified by most farmers. There is no excuse for its costing over two cents per day for each head. A small number may cost more; a large number, less. In my own case, after the first season, it has never required more than 3 hours per day of extra labor, and, for the last two seasons, not over 2½ hours, for 12 head. This includes all the hand labor except plowing the ground. We have shown how the increased value of manure fully compensated for the extra labor, leaving for the profit the saving of land, fences and food, the better condition of the stock and the increase of beef, milk or butter. In fact, I am not able to deny that any one of these seven advantages will, in itself, be sufficient to offset the cost of all extra labor involved.

"Soiling," says Mr. H. Stewart, "is a little more laborious than pasturing, but one dollar spent in extra labor is replaced ten times over in saving of feed, saving of land, and saving of manure. I have found labor very much cheaper than feed." Again he says, "Besides 15 cows, there were 3 horses, 7 heifers, 1 bull (26 head) and some pigs. All the cleaning, feeding and attendance on these animals was done by a boy of 14 years for one year, and the boy had considerable time to spend in field work. * * * The extra labor involved is well repaid by the extra manure made, and the gain from the cattle and the increased fertility of the soil will be clear profit. The *bug-bear* of labor is a phantom. It is imaginary. The need is more for head work than for hand work."

Another writer in the *Country Gentleman*, who has had many years experience in soiling, says; "It requires one man put half of his time cutting, hauling to the barn, and feeding 48 cows, at $1.00 per day." (a trifle over one cent per cow).

I never could see why a farmer should object to extra labor when there is found a profit in it. It is rarely that a man accumulates wealth from the labor of his own hands. The carpenter, blacksmith, shoemaker, or other mechanic who ever becomes well-to-do must owe his prosperity somewhat to the labor of other men's hands. There is a great amount of work to be performed upon a farm that would pay a handsome *profit*. But, as it does not always return to the farmer directly in cash, and sometimes only after an interval of years, he is inclined to apply himself to such work only as puts the "almighty dollar" directly in his pocket. This, I think, is the chief reason why the soiling system is not more generally practised. Many do not like to see a crop of green rye, oats or peas cut down and fed to stock, when, by waiting a few weeks longer, they could harvest it and deliver the grain to market for cash. It has often been remarked to me by visitors at my place, who have witnessed the cutting of a splendid crop of oats or rye as they were just heading out, "What a pity!" It is a greater pity in my estimation, to see a man so short-sighted as to become "penny-wise and pound foolish." Such men try to see how little they can feed and keep their stock alive, begrudging even the insufficient fodder. They go on year after year, plowing wheat after wheat, yearly reducing their stock and the fertility of their soil, and grumbling because "farming don't pay." I have no sympathy with such men. They are unworthy of the name of *farmer*.

LACK OF EXERCISE.

In regard to feeding stock for profit, Mr. Willard says; " The quantity of milk may be increased if certain circumstances and conditions are observed. And first among these conditions is quietness and freedom from anything like labor or extra exertion on the part of the cow. A certain amount of exercise may be needed for health, but all exercise produces a waste of the animal structure, which must be repaired by food. The first office of food is to support respiration and repair the natural waste of the body, and, if the waste is excessive by reason of excessive labor, the food will go first to supply this waste and, after that, for the production of milk. Hence, those who study to get large results from milch cows are careful to *keep the animals as quiet as possible*, avoiding excessive travel or labor, taking care that there be no disturbing cause for excitement, such as fear, anxiety, or solicitude, for these waste food and check the secretion of milk to a much greater extent than most people imagine."

As to keeping the cows confined, however, the soiling system does not require it any more than pasturing. Stock soiled may be fed in large racks in the barnyard, or a small enclosure, where they may have the same liberty as in the field without destroying their food by tramping on it, &c. My first experience was to feed in racks about the yard, but I soon abandoned it, for various reasons:

1st.—The master cows would occupy 10 or 12 feet of a rack, and, all others had to keep off.

2d.—The timid and weak ones did not get their share.

3d.—There was more danger of their hooking each other.

4th.—It required more feed.

5th.—The animals were unprotected from flies and storms.

6th.—The manure was not so good.

All these objections were to be overcome by stabling. The extra labor of cleaning the stables and supplying them with bedding was amply compensated for, because the cows were much more easily and quickly milked, and no chasing about the yard nor clubbing with milk stools, &c. Stabled, the cows could be milked as they should be, at regular intervals, at six o'clock, morning and evening, instead of before sunrise and after sundown, as farmers are obliged to do in fly time, or endanger their sight and tempers. My plan has been to allow them the run of a well littered yard during the night, so that they may enjoy as much exercise in the cool night air as they require.

Mr. Volney Lacy, of Caledonia, N. Y., soils from 40 to 60 head of grade Durhams, which, during the winter, are never allowed to leave their stables, except long enough to drink, and, during the summer, never leave his barnyard. He has cows in the herd 8 and 10 years old, that were raised on the place and have never been out of the barnyard. A thriftier, healthier or more uniform looking herd of cows I never saw. He has a farm of over 200 acres without a rod of interior fence upon it. He has practiced soiling for the last 14 years. I shall always remember my visit to his place with pleasure. *He is a farmer.*

SOILING CROPS.

In regard to the different forage crops that may be used for soiling, as they are numerous and various, they must be selected by the soiler with reference to the nature of his soil, and the condition of his farm and stock. We will notice only those that have come into general use for this purpose.

RYE.

This, by reference to the foregoing tables of the relative value of crops, will be seen to be, for more than one reason, very valuable. We see that, as a green food, it ranks second in the list. There is probably no other plant grown for soiling which furnishes such an abundance of food early in the season. It occupies the ground when no other crop, except wheat, will grow. It is far less sensitive to cold than wheat, and its vegetation is more rapid. It may also be cultivated longer on the same soil than any other crop of cereals, as it is far less exhaustive to the soil. It will produce a fair yield where wheat will not pay the expense of growing. For these reasons it is also one of the best crops to grow as a green manure, as it is equal, ton for ton, to ordinary barnyard manure.

In the soiling system, this is no small item for consideration, as, in the fall of the year, nearly all the ground that will be required for soiling the coming season may be sown to rye and what remains unconsumed in the spring may be plowed under for manure, thus saving drawing, piling and spreading.

Dr. Hamlin, in his valuable work on "Farming with Green Manure," says, "When we compare it (rye) with barnyard manure its greatest value as a green dressing becomes apparent. I have seen fifteen tons per acre growing on the 8th of May, and this was ascertained by careful measurement."

This makes indeed a very cheap fertilizer, viz: seed, $2.00, and interest on the value of the land from October until May (8 months) $4.00, or a total cost of only $6.00 for 15 tons of green manure. The same amount of barnyard manure could not be bought, drawn to the field and spread for less than $20.00. Rye should be fed during the younger stages of its growth. After the heads are formed it soon becomes tough and stock reject it. If cut before heading, on very rich soil it will sprout and grow again, some claim to obtain two or three cuttings, allowing the last one to ripen for seed.

BARLEY.

This is a valuable crop for early spring and late fall sowing, and is quite indispensable to farmers living in a latitude north of the southern boundary line of New York, especially if no preparation has been made the previous fall by sowing rye, or where farmers may wish to pasture during the months of May and June, and soil the remainder of the season. In this latitude (Western New York), it may be sown earlier in the season than any other crop, as it will sprout and grow at a lower temperature. It will also withstand late and early frosts. Mr. A. W. Cheever, of the *New England Farmer*, says, "Two year's experience with barley for cutting in September, October and November, shows that it is very valuable for late fall feeding, as it is not much injured by frosts. Some of my neighbors have been cutting it this season (1879) even after the ground was frozen." For this purpose, the six-rowed barley is said to better withstand the cold than the two-rowed variety. Says Mr. Flint (*Grasses and Forage Plants*, p. 164), "it has passed into a six regular rowed variety, which is a winter grain and endures more severe cold."

CORN.

There is doubtless no other forage crop that will produce more tons of fodder per acre than corn, and this quality strongly recommends it to persons living in a thickly populated district and whose acreage is therefore limited.

On good land, half a square rod of corn, in drills two feet apart, will be found sufficient to support one cow a day (24 hours); and, on land in a high state of cultivation, a quarter of a square rod has been found sufficient. In the first instance, we have, from an acre, 320 days feeding for one cow, and, in the second, enough to support three fullgrown cows 200 days. In selecting a variety, the kind that is found to be the most leafy and, at the

"SOU" FODDER CORN.

same time, not too coarse in the stalk, is generally considered to be the best adapted to the soiling system. Stowell's Evergreen, Blunt's Prolific, Western Dent, and the common field varieties are largely used. Sorghum (sugar cane) is highly prized by some.

It should be sown in drills wide enough to cultivate. When sown broadcast, the leaves, which are the most important part of the plant, stop short of full development, the stalk is weak, and liable to be thrown down by storms and has not the strength to right itself. Sow 2 to 2½ bushels per acre, in drills 2 to 2¼ feet apart. It is hardly necessary to add that the ground should be well manured and cultivated. Mr. Harris Lewis says he has found Stowell's "Evergreen" sweet corn makes the richest milk of all the plants he has tried.

A variety known as the "Sou" Fodder Corn—represented by the accompanying illustration—is very highly recommended by those who have tried it. Possessing as it does, all the requirements most desirable either for soiling or ensilage, it must soon rank as a very valuable variety as its leaves are *very wide* and *very numerous*. The stalk is small, growing upon rich soil, from "12 to 14 feet high," and it is said to contain a large per cent. of saccharine matter. I shall give it a trial.

OATS, OR OATS AND PEAS.

For the production of milk, oats, or oats and peas (⅓ peas,—common field variety), make a valuable crop. As far as my experience goes, I may say, that this is my favorite crop for milch cows.

Sow in the spring, as early as the ground will permit, and begin to cut when the oats are heading out,—the peas will then be leaving blossom and forming pods. It is also an excellent soiling crop for sheep, especially when suckling lambs, also for mares suckling colts. I also feed it quite extensively to work horses, in which case I prefer the plant to be somewhat further advanced in growth. In a recent letter from Mr. Crozier, of Long Island, after mentioning several of the leading crops that he uses for soiling, he says, "I also grow that most valuable crop for soiling, *oats and peas, one of the best crops I can grow.*" Mr. T. Brown, in the *Country Gentleman,* gives it as his experience that oats cut and fed green will produce the most milk of all green crops and will be the greatest profit to the cheese dairy.

CLOVER.

The value of this most nutritious food, is too well known to need any description here. It is one of the easiest crops with which to begin the practice of soiling. If cut just before blossoming, it will furnish three cuttings in a season. I have not used it much in soiling, except for horses and hogs. For them I know of nothing better except tares, or, perhaps, lucern (with the latter I have had no experience). The principal reason why clover is not more extensively used as a soiling crop is, that while it is very valuable, there are other crops used instead which produce two, four or six times as much per acre, and yet are not so valuable for hay. It is much cheaper to cut the feed for 12 or 14 cows from 5 or 6 rods (per day) than to cut it from 10, 20 or 30 rods. "One acre of clover," says Mr. H. Lewis, "will feed a dairy of 45 cows 15 days," and he adds that 3 acres furnishes his herd of 38 cows by soiling 5 weeks. As a green manure, doubtless it has no equal, except perhaps in the cow pea. The action of its roots, which are very numerous, are both mechanical and chemical, loosening the soil and admitting the air, and the decaying roots furnish considerable plant food for the following crop. But one of its most valuable qualities is that, when allowed to grow for soiling crops or hay, it so shades the surface of the soil that it increases its fertility rather than exhausting it, which would not be the case were the land used for pasture. Mr. E. W. Stewart says, " Desiring to know the feeding capacity of an acre of clover * * * I measured off 40 square rods and began feeding it to seven cows and five horses. To my surprise, it fed them 15 days,—equal to feeding one cow 180 days. The two succeeding years I tried the same experiment, feeding only cows, one of which proved equal to feeding one cow 170 days, the other 165 days.

LUCERN.

Where it has a favorable soil, lucern is, in some respects, superior to clover. They grow very much alike, except the lucern grows taller and is a more leafy plant. It lasts much longer than clover, remaining in the soil ten to fifteen years without re-seeding. Those having soil to which it is adapted are loud in its praise as a forage plant. Says Mr. Flint (" Grasses and Forage Plants," p. 193), " But, notwithstanding the great amount of succulent and nutritious forage it produces, its effect is to ameliorate and improve the soil rather than to exhaust it. * * When properly managed, the number of cattle which can be kept in good condition on an acre of lucern during the whole season exceeds belief. It is no sooner mown than it bushes out fresh shoots, and, wonderful as the growth of clover sometimes is in a field that has been lately mown, that of lucern is far more rapid. Lucern will last for many years, shooting its roots, tough and fibrous, downward for nourishment till it is out of the reach of drouth. In the dryest and most sultry weather, when every blade of grass droops for want of moisture, lucern holds up its stem, fresh and green as in the genial spring."

The following is from Mr. John Bruce, from the " Ontario Experimental Farm," Hamilton, Ontario:

" We have had most marked success in growing lucern, in all kinds of weather (hard winters and dry summers) giving 20 tons per acre from four or five cuttings per season.—Broadcast seeding in a free deep soil, clean and in good heart'; 16 pounds per acre sown in spring,—plants last from 6 to 8 years. Give a good top-dressing of well rotted barnyard manure every second year.—It is a good soiling crop; in great favor with cattle and sheep, —is very fattening."

From the late N. Bethel, Thorold : "A portion of my first piece has been down ten years and the other part six years. I cut three tons of hay to the acre last week in June and allowed the second crop to stand for seed. I sold 39 bushels clean seed. This field has never had any kind of manure since it was sown. The other field contains 4½ acres and this is the first year it has been cut. I got about 2½ tons per acre, but it was a very dry season, and on hard clay soil. The second crop produced 25 bushels of seed. I used the cut hay unmixed with bran or other feed for lambing ewes and they gave more milk on it than any feed I have used in my experience of 20 years; it is also very good for milking cows, either as green fodder or hay. The seed can be saved from either the first or second crop. I consider it the best fodder crop I have ever seen, as it stands drouth better than anything I know. I cultivate the same as red clover."

HUNGARIAN GRASS (MILLET).

This is doubtless the most nutritious green forage that is used in soiling cattle, as may be seen by reference to the foregoing tables. As a green manure it also ranks first, containing 20 lbs. of nitrogen and 17 lbs. of potash to the ton. It germinates and grows very rapidly and is said to enured drouth remarkably well, " remaining green even where other vegetation is parched, and if its development is arrested by dry weather, the least rain will restore it." It is a very leafy plant and furnishes the most succulent food, which is highly relished by all kinds of stock. It is said to flourish in somewhat higher and dryer soil than other grasses, but attains its greatest luxuriance in soil of medium constancy and well manured. It is usually sown broadcast, requiring one bushel of seed per acre.

COMMON MILLET.

This is similar to Hungarian grass in regard to cultivation and growth. They are both annual grasses. Of millet, Mr. Flint says: " It is one of the best crops we have for cutting and feeding for soiling purposes, since its yield is large. Its luxuriant leaves are much relished by milch cows and other stock. It requires good soil and is rather an exhaustive crop, but yields a product valuable in proportion to the richness of the soil."

ROOTS.

For late fall feeding, some kind of forage should be provided that will resist early frosts. As before stated, barley or rye may be grown later without injury from frosts, than any other cereals except wheat, but, as in this latitude we sometimes have early falls of snow before winter fully sets in, which would beat down a crop of rye or barley, thus destroying its use as a food, it is best to be provided with some kind of a root crop, such as Beets, Cabbage, Kohl Rabi, &c., for milch cows, and Turnips for fattening animals and sheep. I speak in this way of Turnips, not because they are superior to any of the other roots, but because they cannot be fed to milch cows without injuring the flavor of the milk and butter. Of the turnip, I shall have somewhat to say, under the head of soiling crops for sheep.

BEETS.

As far as my experience extends in growing roots for milch cows and feeding them to obtain butter I have found the " Yellow Ovoid " variety superior to all others. They are easily cultivated and easily harvested, as the principal growth is above ground, as represented in the accompanying illustration. In regard to the method or culture, I need only say, that I have obtained the best results by top-dressing with stable manure after the beets were up, and following immediately with cultivation, working the manure towards and on the rows. The seeding is done with a common field grain

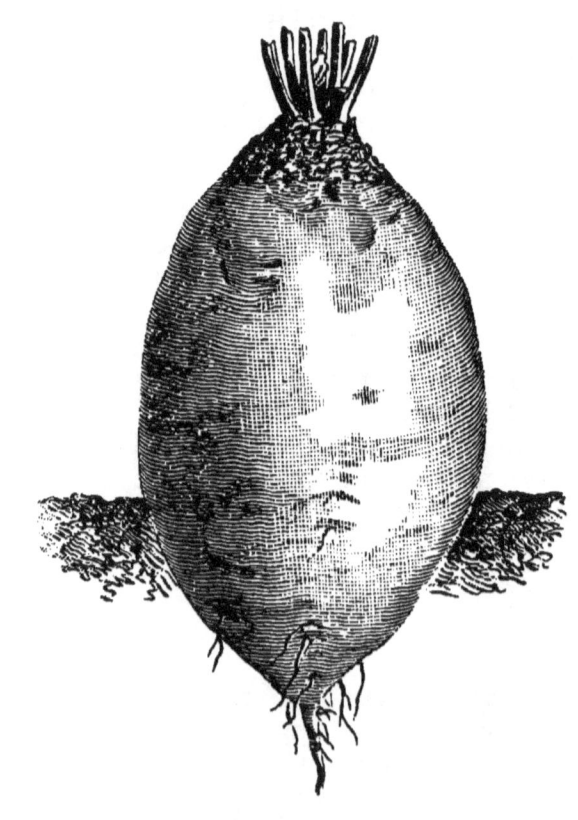

YELLOW OVOID.

drill, in rows 24 inches apart, by closing all the drills except Nos. 2, 5 and 8. In estimating the amount, the drill may be operated upon the barn floor and so rigged as to drop the seed in any desired quantity.

TREE CABBAGE.

CABBAGE.

Besides the ordinary variety of field cabbage used for soiling, the "Tree" cabbage, where it is known, is generally preferred, as its leaves may be picked off several times during the season, growing out again after each picking. Its cultivation in this country is very limited, but an English author speaks very highly of it as a soiling crop. Its cultivation is similar to cabbage, but it doubtless will require more labor to grow it than most American farmers are willing to devote, especially if there is any other forage that will answer the same purpose and require less cultivation. In this respect the Kohl Rabi

KOHL RABI

will be better received. This vegetable as yet is little known in this country. It comes very highly recommended from the German and English farmers, and it is thought by some that it will soon be more extensively cultivated there than the turnip, that is to say, that it will supercede it as a vegetable. It partakes of the nature both of the cabbage and of the turnip. Its cultivation is similar to that of the turnip. It is equally as productive, and may be fed to milch cows, as it gives no unpleasant flavor to the milk as in the case of turnips. It is also said that it will endure very severe frosts.

ROTATION OF SOILING CROPS.

The rotation of soiling crops is very easily followed. The manner of conducting the system of growing and feeding is shown by the following plan, which I have arranged in six steps of growth, one for each of the six months feeding, beginning about the 1st of May, each growth or crop furnishing feed for one month:

1st.—Winter rye sown in the fall, for the next May feeding.

2d.—Barley, Oats, or Oats and Peas for the first spring sowing, (for June feeding).

3d.—Corn fodder, millet, or Hungarian grass, (for July feeding).

4th.—Corn fodder grown on the ground that was occupied by the winter rye, (for August feeding).

5th.—Corn fodder grown on the ground that was occupied by the 2d crop, (for September feeding).

6th.—Barley, rye, or cabbage roots (for fall feeding) grown on the ground that was occupied by the 3d crop, (for October feeding).

If any of the prominent grasses (lucern, clover, &c.) are cultivated, they may take the place of the 2d growth. The ground on which is grown the 1st, 2d and 3d crops, if well manured after each cutting, or in a high state of fertility, will be all the land required for the season, except, perhaps, it may be found in a lower latitude that the 3d crop will be insufficient to carry the stock through until winter; in which case, other land should be provided. The 4th and 5th crops will be off in time to sow to winter rye for the next year's feeding.

HOW TO BEGIN.

In laying out the work, it is simply necessary to know how many head of stock (1,000 pounds each) you desire to soil. On my own farm, I have adopted the following estimate *i. e.;* that a full grown cow will consume per day:

Of lucern, clover, or other grasses, 1 square rod.
Of rye, barley, oats and peas, ¾ square rod.
Of corn fodder, ½ square rod.

Land in a high state of cultivation will require less, poor land more. I would, however, advise a beginner to make a liberal allowance, especially in his first attempt, or until he has become acquainted with his soil and the system. There is no waste in having too much for immediate use, since it may be cut for winter feed, or, better yet, plowed under as a green manure. That the reader may fully understand my meaning, let us suppose that we wish to soil, during the coming season, 12 head of cows. We will begin with the proper spring work, supposing we have sown sufficient rye the previous fall to sustain our herd through the month of May (1st crop).

SECOND (JUNE) CROPS.

The plowing should be done as soon as the ground will permit. The land should first have a heavy dressing of manure, and, after plowing, it may be also top-dressed. All the land that may be required for the 2d and 3d crops (June and July feedings) may be plowed at once.

As already described, the succeeding crops (August, September and October) may be grown on the same ground as the first three. The next thing in order is to ascertain how much land will be required. At ¼ square rod per day for each cow, we shall need for the second (June) crop (barley, oats and peas), 270 square rods; to which we will add 10 square rods for waste, &c., making 280 square rods. In plowing for both June and July crops, we plow, of course, double this area. The 4th (August) crop will come, as before stated, on the ground now occupied by the first crop (winter rye).

Now that the ground is all plowed, the man who has charge of the cutting and feeding proceeds as follows: every Saturday afternoon, with two horses he fits the ground, and with the grain drill, sows, only the amount of ground necessary to keep the 12 head *one week*. This makes four sowings, of 70 square rods each, for each month's crops, and thus each sowing in rotation comes to maturity one week later than the previous one. One week is about as long as most forage crops are at a proper stage to cut. If, for instance, two or three week's feeding is sown at a time, the soiler must begin cutting before it has reached its best state (*i. e.*, in blossom), or else it will not all be consumed until some of it has passed its most succulent state and become so tough as to be rejected by the stock, the soiler may thus be disappointed and led to condemn the system. For the lack of knowing this, I have no doubt that more than one faint hearted man has been discouraged in attempting to soil his stock.

It has been my practice to sow barley for the first week's sowing and to follow it with three weekly sowings of oats and peas. This constitutes the second month's crop. The rye will last until about the 1st of June when the first cutting of barley will be ready, after which, each week's sowing follows in succession. The sowing must therefore be continued every week until each month's crop is provided, and will require, for 12 head, about two or three hours of one day in each week.

THIRD (JULY) CROP.

If the grasses are used, provide at least 1 square rod per day for each head; if corn fodder, ½ square rod. Say we use the latter, and that we use 200 square rods for the whole month's crop. Four sowings would therefore require the fitting and seed for 50 square rods each week. For reasons already given, the seed should be put in with drills and cultivated. The seeding may be done with a common grain drill. With a nine-tooth drill, let the cups of teeth Nos. 1 and 3 discharge into No. 2; 4 and 6, into 5; and 7 and 9, into 8. This will make the rows 24 inches apart, the wheel tracks serving as a guide on return bouts.

FOURTH (AUGUST) CROP.

This crop is usually corn fodder, which thrives well in hot weather. It may follow on the land lately cleared of the winter rye, after manuring; but, if the land is not in condition, it must be grown elsewhere.

FIFTH (SEPTEMBER) CROP.

This follows on the land where we sowed barley and oats for the June feeding.

SIXTH (OCTOBER) CROP.

This should consist largely of barley or rye, or both, sown, in the manner above described, where we had the third (July) crop.

Cabbage and roots for November should have been sown early, on ground especially fitted for them. But many farmers allow their cows to run out during the latter part of the season (November) on corn stubble or such land as they have suitable for such uses. In September, there should be the usual number of sowings of rye for the first month's feeding in the following spring, which should be top-dressed during the winter, when the ground is frozen, with fine stable manure.

This completes the season. It seems as if I had used a great many words in describing the rotation of the crops and the manner of growing them. If I am at fault in this respect, I hope the reader will attribute it to my desire to be clearly understood.

In this connection, I add the following extracts from a letter from Mr. Charles W. Wolcott:

Blue Hill Farm, Canton, Mass., June 11th, 1881.
F. S. PEER, Esq.,

DEAR SIR:—I have yours of the 4th, and note the inquiries. Our practice has been to feed on winter rye first then oats,—next, spring rye,—next, millet (the golden) grown on the winter rye land. *Sweet* fodder corn (Stowell's evergreen) grown on oat land, southern white fodder corn sown in drills on oat land and spring rye land, and lastly, *barley* grown on the land formerly occupied by winter rye and lastly by golden millet. This gives a good rotation for feeding, and with us always has worked well. * * * * Respecting the value of manure saved by soiling, my judgment is that *all* that is *made* is *saved*, for I do not believe that the manure dropped in pasture enriches the soil at all, it being mostly dried up into an almost insoluble cake.

The care of my stock (now 48 head of milch cows) devolves on one man who feeds, cleans and waters them in the barn—two men help him milk. One man and one horse draw the green fodder in less than half a day. We feed three times a day in the stanchions, where the cows stay except when they are turned out in the yard once a week for an hour if it is cool, but never if it is very hot. They much prefer the barn to the yard. Their health is always good and they are thrifty. The quality of milk is about the same with me the year round. The quantity is larger with me in the soiling season than my neighbors average.

To conclude, I will say that I cannot see that I can afford to pasture my stock, as I haven't made enough money yet to be ready to throw it away.

Yours, Respectfully,

CHAS. W. WOLCOTT.

CUTTING AND FEEDING.

CUTTING.

Where the farmer has but a few head to soil, a large wheelbarrow, scythe, cradle, rake, fork and corn cutter will constitute the necessary outfit to conduct the system, so far as cutting and feeding are concerned. My own experience in soiling 12 to 14 head of milch cows and 4 horses may be briefly stated as follows: The cutting is all done with a stoutly built self-rake reaper. We began by using a scythe. The second year we cut with a mowing machine, but this had some disadvantages over the scythe, for the crop had to be raked together either by hand or by horse rake. I tried the latter, but it was not stout enough to rake the heavy green fodder in any better shape than it was left by the scythe. It required considerable hand raking, either way. This I found to be the greatest objection to the whole system. Not that it required so much time, but that to go back to gathering crops with hand rake and fork, after a few years with mowing machines and hay rakes, didn't seem like progressing. We used the scythe until the new reaper (D. M. Osborne's No. 3, manufactured at Auburn, N. Y.) made its appearance on the farm, taking the place of scythe, mowing machine, hand rake and horse rake; in fact, for ease and economy of labor, this machine proved to be just the thing. Monday morning, for instance, the farm team is attached and cuts, in twenty or thirty minutes, enough feed to supply my entire stock for two or three days, leaving it in the very best possible shape to gather, where it may wilt without drying out, and the least exposed to sun and weather. I cannot recommend too highly or encourage too strongly this mode of cutting over every other. This is the third season that I have used this machine, cutting very heavy crops of oats and peas, and last season, besides all our soiling crops, 5½ acres of Western Dent corn, 7 to 9 feet tall, and producing 30 tons of ensilage fodder per acre. Of this subject, I shall have somewhat to say under "Winter Soiling." The only break that has ever happened to the reaper while cutting green feed was a casting on the "grain wheel" outside of the table. I speak of this because, at first, one might be timid about using a reaper for such work.

DRAWING.

A one-horse lumber wagon with wide tire (2½ inch) and low wheels (half truck) will be found of great service for drawing in the green fodder, drawing out manure on plowed ground for top-dressing, and drawing out muck, &c. It should have a double box, the upper part projecting on a bevel so as to cover the wheels as here represented.

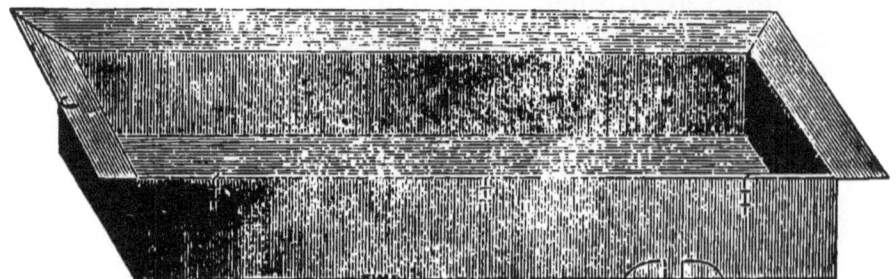

FEEDING.

There are two methods of feeding stock under the soiling system, viz: By allowing the stock to run in a small enclosure or barnyard, where they may be fed in racks; and by feeding them in their stalls, allowing them such exercise as the owner may deem necessary to promote and maintain their health.

The latter method has many advantages over the former, some of which have been mentioned under the head of Advantages, *i. e.*; the saving of feed and manure, and the greater comfort of the stock. Lastly, but not least, may be mentioned the advantage of having the cows in their stalls at milking time, instead of being scattered about the yard, driving each other about, which is very annoying and often results in a clubbing of the cow and the loss of a pail of milk.

CAUTION IN FEEDING.

There is much more danger of a beginner's feeding too much than not enough. A cow, with more fodder placed before her than she can eat up clean, breathes upon it and then will go hungry before she will eat it. This, of course, causes shrinkage in milk and flesh, and I have no doubt that for this reason the soiling system has been condemned by some, who supposed that a cow could not be hungry with a whole rack full of feed before her which she refused to touch.

MANNER OF FEEDING.

Experience has taught me that, to produce the best results from milch cows, they should be fed four or five times a day, and each time only the amount which they will consume before the next feeding. Mr. Quincy recommends six feedings; but five, in my experience, have given equally good results and are more convenient. To think of feeding five times a day may seem like a great task, but, by systematizing the work, it will be found not nearly as bad as one may imagine.

Let us begin in the morning and go through the entire day's work of soiling 14 head of cows, feeding them five times, twice in the morning, once at noon, and twice at night, as follows; at 5 and 8 o'clock A. M., 12 M., and at 4 and 7 P. M. We have enough feed brought in the night before to supply us with the first, or 5 o'clock, feeding, which is, at time of feeding, placed in the mangers and the cows let in from the yard where they have spent the night. The farm team is then cared for. By this time breakfast is ready, after which (6 o'clock) milking begins. In the meantime, the calves and pigs are attended to, the farm team has been attached to the reaper, and, having cut sufficient fodder for two or three day's feeding, has gone to the regular field work. After the feeding of the calves and pigs just referred to, the man, with the wagon, draws from the field, in one load, feed for the 8 o'clock, noon, and 4 o'clock feedings,

After all other chores are done, which takes him till about half-past seven, the cows are given their 8 o'clock feeding, and the man is at liberty until noon. Just before going to dinner he gives the cows their noon feeding, which takes him about

CUTTING AND FEEDING.

five minutes. After dinner he is at liberty until half-past four when the cows are again fed (what we call the four o'clock feeding). Supper at five, after which another load of fodder is drawn from the field, sufficient for the 7 o'clock feeding and the 5 o'clock morning feeding. At 6 o'clock the cows are milked, and the calves and pigs fed, which brings us to the 7 o'clock feeding. This is quickly done, and ends the day with the exception of turning out the cows at 8 o'clock. During the time we have given to feeding the 14 cows, 4 horses and 2 colts have been provided with feed.

In relating, as I have, my own experience in conducting the soiling system, I am well aware that it establishes no rule. It might not suit another's case in every respect. I hope, however, that it will give my readers a correct knowledge of the general principles of soiling, so that those who wish to adopt it may have at least a guide, if not an absolute rule.

SOILING CATTLE.

In the construction of cow stables, there are two points which should be well understood, and the first one of these is the necessity of good and perfect ventilation. This is very easily secured and should in no wise be neglected, especially if stock are to be stall-fed. Either in summer or in winter, pure air is indispensable to health. A stable constructed so as to be warm in winter, will be cool in summer. The following diagram illustrates the principle of ventilation:

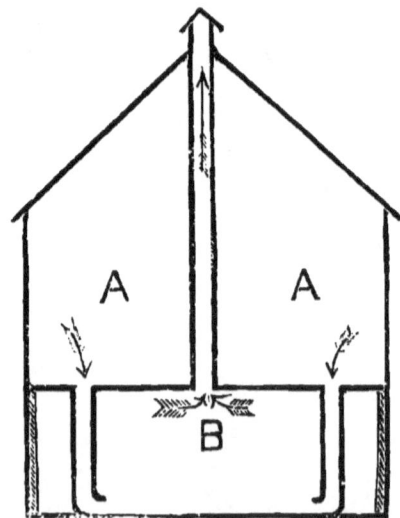

As the warmer and impure air from the breathing cows and the odors of the stable rise to the ceiling of the basement stable B, it finds an easy exit upward and through the roof of the building by the air duct represented. At the same time, the circuit is completed by the introduction of cooler and purer air from the floor of the upper apartment, A A, through the conduits represented, to the floor of the basement stables.

The second item of importance is the saving of liquid manure, the value of which has been already discussed. Figure 2 gives a sectional view of a double row of stalls facing each other, in a basement 30 feet wide.

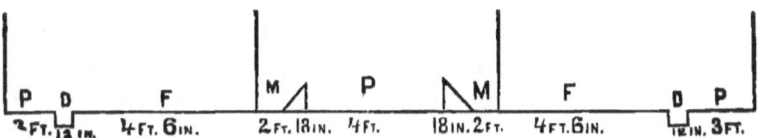

The space in the center, marked P, is the alley through which the feed is carried and deposited in the mangers M M. F F are the floors on which the cows stand. The drops (D D) behind the cows are made water tight and may be partly filled with absorbents. The spaces P P, at the extreme ends of the diagram are passage ways behind the cows.

I prefer a deep drop behind the cows, to a shallow wide one, for these reasons. It takes up less room; a cow can more easily step across and is less liable to slip in passing; it is more easily cleaned out, being but the width of a large scoop shovel; and, being deeper, the manure drops below the surface of the floor on which the cows stand, and, if cleaned out daily, there will not enough accumulate to touch the body of the cow when she lies down. I have had a solid white cow in my stable, summer and winter, for the last three years, and I never remember seeing a manure stain on her flank, legs or udder. The same is true of all the cows. I use fine shavings for bedding and as an absorbent. The cows work enough back into the drop or trench with their feet to absorb the urine. When straw is used for bedding, a little dry muck, sand, cobs, etc., should be sprinkled in the bottom of the trench, as straw itself is not a very good absorbent. If cut cornstalks are fed, the butts left by the cows may be thrown under them; they make a good absorbent, but should, of course, be cut. Muck and sand are, perhaps, the best absorbents, on account of being of greater value before being used than the others I have named; but, where it is the practice to furnish cows with bedding, something that will answer for both bedding and absorbent will doubtless be the most desirable.

FASTENINGS.

There are many styles of cattle fastenings, the ones in most general use, being ropes, chains, and stanchions. These are too well known to need description. The foregoing illustrates the latest improved method of fastening by stanchions.

For some reasons stanchions are the best method of fastening. The objections generally made to them are that the animal has not sufficient liberty with her head. The accompanying illustration shows how this objection is obviated. This style of stanchion is manufactured and sold by the patentees, M. H. Barnard & Co., Forestville, Conn.

SOILING HORSES.

My own experience being limited to the soiling of the farm team only, and not extending to the soiling of breeding mares, I quote in full the following article on the subject from the pen of a well-known author and practical farmer, as it appeared in the *Live Stock Journal :*

"This class of stock is thought by many to be quite unadapted to the soiling system, especially colts, as they require exercise to develop the muscular power, and soiling is thought to require too close confinement. This arises from a misconception of the flexibility of this system. Soiling does not necessarily require the confinement of animals any more than pasturing. It is true that pasturing furnishes larger fields to range in; but nearly every farm can devote a lane running to the wood lot as space to exercise in. This lane is necessary for the convenience of the farm, and generally furnishes a road to the different parts of the tillable land and meadow. This will furnish the colts abundant room to make trials of speed, and afford all the exercise required to develop muscle. This run-way is easily fenced so substantially as wholly to prevent the colts from jumping, and thus becoming troublesome. I have raised a dozen colts in this way, and found them to develop in every respect as well as those pastured. That colts may be as little confined as possible, racks may be arranged under a shed, into which the soiling food may be placed, and the colts have access to it at all times. We found this plan to work well with brood mares and their foals. Having the food of the mares wholly under control, their production of milk will be more uniform, and the growth of the foals much better, than on pasture. The dam requires full feeding upon appropriate food, and this may always be given in soiling, as any defect in the succulence and nutrition of the grasses or other soiling food may be supplemented with middlings, oil-meal and oats. The foals are also constantly under the eye of the feeder, easily become accustomed to handling, and may be taught to take other food at a younger age. Early familiarity with the attendant and docility are not only favorable to the foal's progress in development, but to its easy management at the training age. The vigorous, steady, and healthy growth of colts is most essential to their future value as serviceable animals, and, therefore to the profit of the breeder. Soiling offers the most complete control over the food and management of colts; and, therefore, under this system they may be grown with much more uniform success, and, on land worth fifty or more dollars per acre, much cheaper than by pasturing. As I have shown in other articles, the foal responds more quickly to the use of cow's milk than any other food after weaning, and this may be skimmed milk, after teaching it first to drink new milk. The colt being under attention in soiling, this extra food may be given with very little labor. From considerable experience, I consider the soiling system as well adapted to the raising of horses in all stages, from the suckling colt to the mature horse."

SOILING SHEEP.

The advantages of soiling sheep in this country are becoming more apparent every year. "The flesh and wool of sheep," says Mr. Stewart, "are but the products of the soil, and contain nothing but what has existed in the plants which the sheep have consumed." No farmer who has ever bred sheep for mutton needs to be told the necessity of supplying an abundance of succulent food for his lambs until they have reached maturity. A lamb that has been stinted by want of proper nourishment or from sickness can never be fattened as profitably as one whose growth has never been checked. The English farmers not only know this, but take every precaution to prevent it, and to this it is mainly due that they are enabled to export to this country, yearly, many thousand dollars worth of sheep, while American farmers might breed as good at home, if they would. For many years we sent to them for stock animals to improve our herds, until Americans learned the art of feeding, and now we are breeding cattle as fine as theirs, and, during the last few years, have sent some valuable sires to England.

But, in regard to sheep, we have yet much to learn. I mean we have to put into practice what we already know, but, for some reason, fail to appreciate its importance. There is not a farmer in America who will not say that "it costs no more to keep a good sheep than a poor one"; but not one in a hundred puts the statement to proof by practice. The English farmer makes no secret of how he produces a flock of sheep that average 200 lbs. each, and from 12 to 20 lbs. of beautiful wool.

It is all explained in one word,—*feed*. Not grain so much as a *never-ceasing supply of rich, nutritious forage* which keeps their stock growing constantly, throughout the entire year. To accomplish this, they have adopted a regular system of soiling, there known as folding or hurdling.

I am told by an importer of English stock that, as a general thing, the English feed less grain than we do. Again, it is very important to the wool grower that his flock should have an abundance of food throughout the entire year. Whenever the pastures fail, the growth of wool is checked, and, if the sheep be afterwards well fed, there will be found at shearing time a weak place in the wool corresponding to the time in its growth when the feed was insufficient. Wool, like milk from our cows, is produced in proportion to the amount of food consumed above that required to support life. Therefore, the want of a proper amount of food is first noticed in the wool, and here is where many farmers are deceived. Their sheep look to be in passable condition and they are satisfied; but the sheep are not growing a profitable amount of wool, as they would if supplied with all they could eat. Says Mr. Miles, "The great development in fattening quality and early maturity * * * has been secured by a liberal supply of nutritious food during the period of growth."

Mr. Youatt, an English author, says: "It is of the *utmost importance* that the ewes should have abundant food in order to produce a flow of nutritious milk while they are suckling, and that the lambs should have plenty of good pasture or other succulent green food when they are weaned."

Speaking of the Lincoln breed of sheep, Mr. Stewart says, "In connection with a system of farming in which heavy crops of roots and green fodder were the chief production, this improved breed became fixed in its character as the heaviest producers of mutton and wool in the world."

During the early part of the season, when vegetation is putting forth vigorously, they do very well in pasture, but, by the time they have overcome the effects of winter, the pasture begins to fail,—just when the lambs are requiring the greatest amount of milk. Then it is that the dams should have better feed than at any other time in in the year. There is no other time when extra food is so much needed. The system of the dam herself must be kept up, a large lamb, often two, derive sustenance through her, and the farmer also expects her, at the same time, to be growing wool. If she is ill provided with the best of food to produce milk, wool and flesh, the wool is first affected, then her off-spring—reaching maturity late, sometimes never; her own body becomes a ready prey to disease, and she goes into winter quarters poor. A few years of such a life hangs her hide upon the fence and gives her carcass to the crows.

There are many farmers keeping sheep who have no interest in their improvement for the reason that every two or three yeare the rotation of the fields shortens the supply of pasture and the flock goes to the butcher. They pick up a few culls after a year and begin another flock which in turn follows the course of the first. The farmer has no object in selecting a good sire as a means of improving, because he doesn't know but what he will have to dispose of his flock another year, if he should be likely to loose a feeding or be short of pasture.

There is probably no source of easier profit on

the farm than a flock of *well cared for* sheep. Manure made from them is richer in nitrogen and potash than that from any other animal,—not excepting the hog and the hen. Their wool and lambs are in the market just when the farmer has the least to sell, they require little care compared with cows and horses, and increase more rapidly. In fact, to deprive a farm of a flock of *good* sheep is to rob it of one of its most pleasing and profitable attractions. There is a way in which they may be supplied with food, rich and succulent, when they most require it, a way in which the lambs may be made to grow continually from birth and be early brought to full maturity, a way in which the farmer can produce the greatest amount of wool superior in quality, manure unequaled in value, and make himself the sole possessor of a beautiful flock of sheep,—and that is by *soiling*.

SOILING CROPS FOR SHEEP.

In selecting crops for sheep, we must remember that the forage should be finer than for cattle or horses. The most prominent are tares (vetches), rape, turnips, lucern and clover (early cut), oats and peas, and some of the small varieties of corn fodder.

TARES.

Spring and winter tares are largely sown in England for soiling sheep, cattle and horses. All stock are exceedingly fond of them. My experience in feeding them is very satisfactory. I have never undertaken to cultivate the winter variety. Spring tares are usually sown in March or April. They are very much like the common field pea, except that the stalks and leaves are finer,—a very vigorously growing plant, highly relished by sheep and lambs. The blossom and pod are similar to those of the pea. A small quantity of oats, barley or rye should be sown with them as a support, otherwise they are apt to lodge, after the manner of peas, which materially lessens their value. They may be sown with a grain drill or broadcast.

An English writer says, "Sheep may be fattened upon them, the milk of cows is enriched and increased by them, and they are extensively employed in feeding horses. They do not require a rich soil."

RAPE.

Rape looks and tastes like turnip tops, but has roots similar to those of grain and grasses; it is, however, of much finer growth. The seeds look like those of the turnip. It grows from 10 to 15 inches high. Although it does not produce as much food to the acre, it is the most nutritious forage plant and is equaled by no other vegetable, as may be seen by the foregoing tables. Its culture is similar to that of the turnip and will sustain about the same number of animals per acre, and may be sown later in the season. It is said to "require rich ground where cultivated for seed, but large quantities are grown with advantage, for feed only, on very poor land." As a food for young lambs, it has no superior. A small patch may be grown in one corner of the pasture, or adjoining it or the place where the ewes are confined, with a lamb creep,—a hole in the fence large enough to admit the lambs and exclude the sheep, with a roller at the top and sides to prevent tearing the wool, as shown in the following illustration.

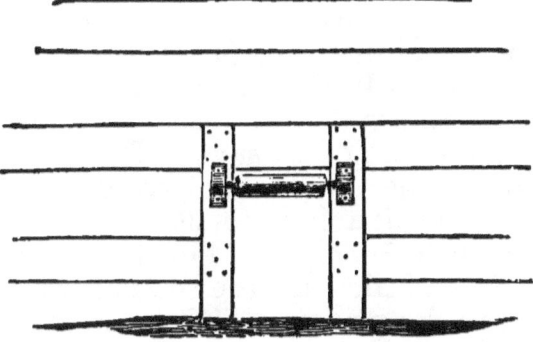

The lambs will soon learn to run in and feed, as they are exceedingly fond of the plant. It requires about two pecks of seed per acre, which should be sown in July for fall feeding. If intended to be fed to grown sheep, it should be cut and fed to them in racks; otherwise, they destroy much of it. Lambs may be allowed to pasture upon it, as they are light in weight, and, if unaccompanied by their dams, only stay in the enclosure while feeding. The high feeding value of this plant strongly recommends it to farmers raising early market lambs. In this case it should be sown earlier.

TURNIPS.

The turnip in England has become a regular rotation crop, and takes the place of corn in this country, i. e.; 1st, turnips, 2d, barley, 3d, wheat, 4th, grass or pasture. The varieties most used for feeding stock are the White Norfolk, Yellow Aberdeen, Swedish, and Dale's Hybrid, " which latter is a hardy, succulent vegetable, much relished by stock, and in no respect injured by the severest winter." It is sometimes sown broadcast, but is found to pay better when sown in drills and cultivated. Mr. Youatt says, " There is no vegetable within the range of agricultural produce that yields so valuable a portion of nutritious food as the Swedish turnip." Turnips may be sown from the last of May till the second week in July.

These are the principal soiling crops for sheep, in connection with the other forage crops which have been considered under the general head fo

WHITE NORFOLK.

soiling crops, especially "Kohl Rabi" and "Tree Cabbage,"

MANNER OF SOILING SHEEP.

We will consider briefly the methods adopted for feeding sheep by the soiling system. If moved about from field to field by the rotation of crops, they may be supplied with any of the soiling crops just mentioned, by fencing off a portion of the field in which they are pastured and devoting that portion to the growth of soiling crops; or a small portion of an adjoining field may be used for that purpose. In either case, the several crops should be sown or planted in four rows parallel with the division fence, the crop for the first feeding being nearest the fence. A movable rack, just in the pasture, will serve to hold the feed as it is cut.

Each row is intended to supply food for one month, beginning about the first of July on the first row, cutting with scythe or cradle and throwing the cutting over the fence into the rack, which may be moved along the line of the fence as the cutting progresses. By the time the first row is consumed, the second should be ready for cutting, which may be done in a direction opposite that of the first, following back with the rack. The first crop next to the dividing fence may be oats and peas; the second, a small variety of sweet corn; third, the same, or tares; fourth, tares. After the first and second rows have been cut, the ground which was occupied by them may be top-dressed and cultivated in, or plowed shallow, and sown to turnips.

In estimating the amount of ground necessary to supply a flock with forage, we apply the same rule as given for calculating the amount required to supply 1000 lbs. (or a full grown cow). Thus, sheep averaging 100 lbs., would require, each, one-tenth of that necessary for a cow, or, of oats and peas, one-tenth of ¾ square rod per day. This estimate for sheep in the plan of feeding above described may be reduced to at least ½ square rod per day for every 1000 lbs., as the sheep will obtain part of their feed from the pasture; but this part will, of course, depend upon the size of the pasture and the fertility of its soil. My own experience by soiling in this manner, was in an old orchard containing 5 acres, one of which was fenced off as above described. With this four acres of pasture, and one devoted to soiling crops, I have kept 24 head of large, long-wooled sheep, and 22 lambs, (fully equal to 5 head of 1000 lbs. each) during the season. This leads me to say that, as a rule, for every 1000 lbs., it will require one acre of land, one-fifth of which should be devoted to soiling crops. I feel safe in saying that the five acres, with one devoted to soiling crops, were equal to ten pastured, or, that one acre soiled is equal to five pastured. The variety of the feed and the shade made the sheep contented, and, better still, they had all they could or would eat.

The following comparison of the amount of wool taken from the same sheep after a year of pasturing and after a year of soiling, shows the effect of their having an abundance of food during the entire season, that there may be no check in the growth of the wool:

1878, 30 head pastured, sheared in 1879, 280 lbs.
1879, 28 " soiled, " " 1880, 330 "
1880, 37 " " " " 1881, 550 "

Those clipped in 1880 were wintered on ensilage and bean straw, as I had no hay. In every other respect they were cared for as in the years previous.

I do not attribute the increase entirely to soiling. When fed in the orchard referred to, the sheep were protected by its shade from the heat of the sun, and such protection has, in my opinion, far more to do with the results of feeding than most people are inclined to admit. The lamb creep should also come in for no small share of credit. My lambs, during the years '80 and '81, weaned and weighed July 1st, have averaged 91 lbs., at an average age of 4 months, many of the single lambs weighing as many pounds as they were days old; a large majority, however, being twin lambs, reduced the average. And, lastly, but not least, the difference should, in some degree, be attributed to improvement of the stock by sires superior in the quality and quantity of their wool. Nevertheless, I believe, that to the *influence of food* (forage) should be credited far the largest part of the result.

FEEDING.

The feeding racks are filled three times a day, morning, noon and night, and this may be done by a boy. No more should be fed at a time than the sheep will eat, and, should there be any left in the racks, it should be removed before fresh feed is added. The shepherd will soon learn the wants of his flock. Another method of feeding is that of folding the sheep upon the soiling crops instead of cutting them. Formerly (in England) this was the custom, but lately they have more generally adopted the practice of cutting and feeding in racks.

MANNER OF SOILING SHEEP.

The best plan of feeding by portable hurdles, is illustrated as follows:

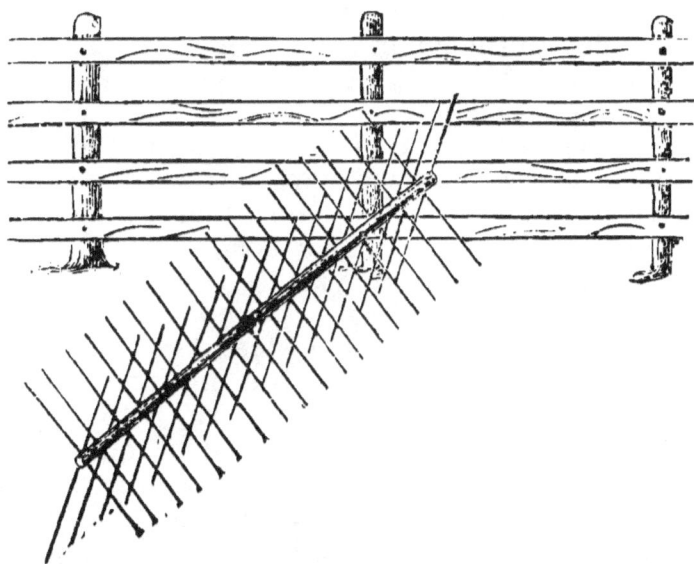

These are made by driving sticks 6 or 7 feet long through a piece of wood 4 inches square and 12 feet long, so that each row will be at right angles to the other and the sticks 7 or 8 inches apart, so as to admit the head and neck of the sheep only.

These racks may be rolled over and over across the field. The sheep are thus provided with fresh food daily, without trampling it down. The inclosure must necessarily change with the movement of of the hurdles, and, therefore, a few panels of portable fence, light and durable are indispensable.

A fence made of cocoanut or hempen cord is often used, which is made in various lengths. Stakes are driven into the ground, on which are hooks for the support of the netting. The netting costs (in England) $9.00 per hundred yards. Mr.

Stewart says of it, " At this price, it could be imported with profit, and probably cheaper than it could be manufactured here."

The following is an extract from a letter lately received from Mr. Charles Barton, Fyfield, England, January 18th, 1882; "I use two kinds of portable fencing, one made of boards, in length 7 feet. The other is a galvanized wire netting, and is made in fifty-yard lengths, which one man can roll up and handle with ease. The cost varies from six pence to ten pence per yard. This I find very handy for summer folding. I keep from 1,200 to 1,400 sheep. * * * The feeding sheep have their roots cut for them and put in troughs with hay and a portion of cake and corn, ½ to 1 pound per head, per day, until fit for the butcher. The ewes with lambs are folded in the amount that we think will be sufficient for them for a day, and their lambs allowed to run forward through the hurdles which are wide enough to let the lambs through, but not the ewes. This is followed until vetches (tares), rape and turnips are fit."

Mr. Barton is a noted breeder of Cotswold sheep and exports yearly some of the finest specimens of that breed coming to Canada and the United States. The following figure represents a steel wire netting which may be used as a portable fence as above described.

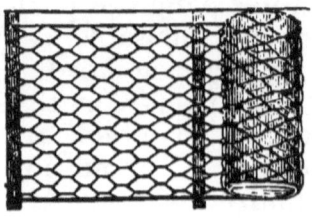
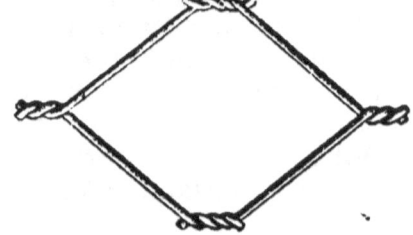

It is furnished either painted or galvanized, in any desired length, for from $1.00 to $2.00 per rod, according to the width, which varies from 3 to 6 feet. It is manufactured and sold by Sedgwick Bros., Richmond, Ind. The illustration represents the meshes of the netting. They are made in two sizes, 4 by 7 inches and 5 by 8 inches.

Another method of feeding is practiced to some extent in this country, where the farmer is so fortunate as to have a permanent pasture on that portion of his farm best adapted to sheep, the size of the field depending on the number of sheep, allowing one acre for every 1000 lbs. A permanent pasture is an excellent feature on any farm. One acre, generally speaking, is worth two or three of newly seeded, by having a field, properly located, seeded with a large variety of grasses, some coming early to maturity, some late, some that thrive best in hot weather, others that do well when cooler, some that grow thickly, to make a heavy sward, and others that send their roots far down beyond the effect of drouth. The following varieties are none too many to make a good pasture. The proportion given is that for one acre.

Sweet-scented Vernal, flowering in April and May,		-	4
Orchard grass,	" " " " "	- -	6
Sheep's Fescue,	" " May and June,	-	3
Kentucky Blue Grass,	" " " " "	- -	4
Italian Rye Grass,	" " June,	- - -	4
Red Top,	" " June and July,	- -	4
Timothy,	" " " " "	- -	4
English Rye Grass,	" " July,	- - -	6
White Clover,	" from May to September,		5
			40

These varieties, flowering as they do, at different times during the season, make them very desirable, besides they are all highly relished by sheep. The seed for an acre will cost from $6.00 to $8.00, which is a trifle more than timothy and clover, and when a pasture of this kind is once established, it becomes very valuable, repaying many times the extra cost of the seed. All of the above varieties may be obtained of Messrs. John Bruce & Co., of Hamilton, Ont., a reliable house.

With such a permanent pasture, the method of growing and feeding the crops above referred to may be illustrated as follows;

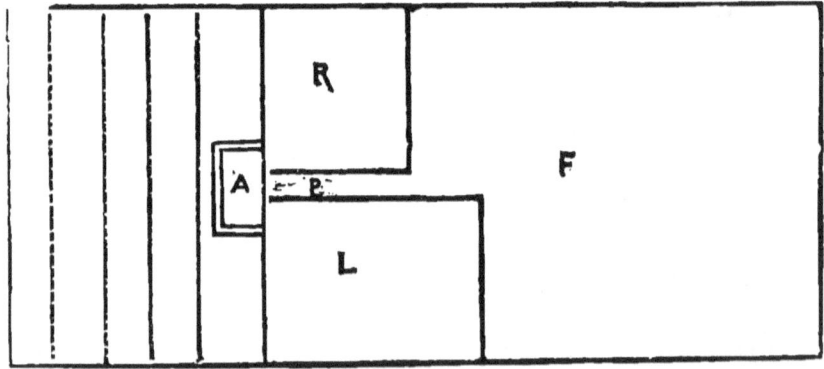

F, L, and R comprise the permanent pasture; A is the feeding shed. The dotted lines at the left show where the soiling crops are grown. R and L are enclosures by which the rams and lambs may be respectively separated from the rest of the flock and confined by portable fences within the pasture, so that they can yet be fed on green fodder in the shed.

The following shows the elevation of the shed:

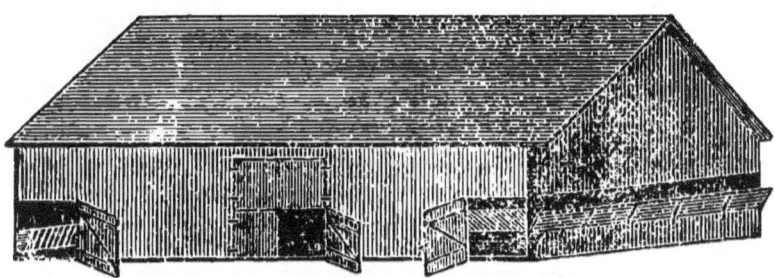

It stands on the ground devoted to soiling crops. The front opens to the pasture with the other three sides within the enclosure devoted to soiling crops, so that the shepherd or soiler may drive on the three sides of the building, putting the feed through into the racks from the wagon without disturbing the sheep by going or driving among them. There are no gates to open and shut, nor sheep in the way to bother about feeding.

I have been experimenting by feeding in a small shed of rough boards during the past season, instead of the movable rack first mentioned, and am so well pleased with it that I propose to enlarge it during the coming season. I like it better because the sheep like it better. They remain inside the greater part of the day and eat considerably more than if in the field, where they lie under the fence most of the time, enriching ground that cannot be cultivated.

This plan requires no more labor on the part of the soiler than when the sheep are fed with the movable racks, and I find that less food is wasted. The value of the manure thus made is also in its favor. There may be a small pen with a lamb creep, where they may help themselves to bran or ground oats as soon as weaned. They may be fed separately on tares, rape, cabbage or turnips, and the rams and ewes fed on coarse feed. In fact the shepherd has perfect control over the feeding and can mete it out as his judgment dictates.

Where a rotation of crops is considered preferable, the building or shed may be situated at the adjoining corners of four fields, as follows:

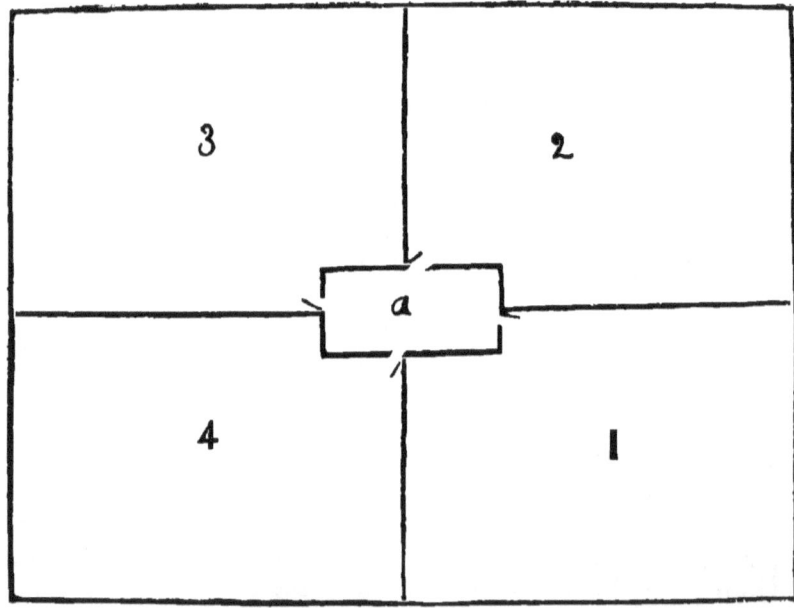

To illustrate: the first year 1 and 2 may be pastured, 3 devoted to soiling crops, and seeded, and 4 for roots. The second year, pasture 2 and 3, devote 4 to soiling crops and seed, and use 1 for roots, and so on around. This plan would doubtless require more land than the other. The choice should depend on the nature and condition of the soil. I have never tried this plan, and so do not speak from experience. I must not neglect to say that the sheep should be supplied with fresh water daily. The idea that a sheep does not require water is simply an excuse for not supplying it. They do need it and should have it. A sheep never cares to drink much at a time, but likes to take a sip quite often. I have found it more profitable to indulge the wants of my stock than my own. They should have salt always before them, in boxes where they can help themselves. I have no doubt that many sheep are injured by eating too much when they are not regularly fed. This may seem to some farmers like "puttering work," but it is strict attention to the details of business that brings success. One of Wellington's friends said to him, on reading the records of his India campaign, "It seems to me, Duke, that your *chief business* in India was to procure rice and bullocks." "And so it was," he replied," "for, if I had rice and bullocks, I had men, and if I had men, I knew I could conquer the enemy." The same principle is illustrated in the soiling system. First secure plenty of food, and you may be sure of stock to eat it. If you supply good food, you may expect good stock; if a large quantity, a large number; if a large number, you may have a rich soil; if a rich soil, you may become a rich man.

Experience of Mr. Charles Francis.

Middleport, N. Y.

Mr. F. S, Peer, Esq., E. Palmyra, N. Y.

My Dear Sir:—In reply to your request as to the "methods I have adopted in growing, handling and feeding soiling crops," I will reply by saying that during the present season, I am soiling eighteen head of milch cows, and must say, that every year's experience with the soiling system, leads to a firmer conviction that it is the most economical and profitable way of feeding farm stock.

My farm fences were in very bad condition when I bought it. I took possession in the spring of 1878. I spent considerable money and all the time we had the first season in building fences. But one cannot fence a farm of two hundred and forty acres all at once. An early spring soon drove us to the fields, with the amount of fencing that I had intended to build, not half complete.

The result was that we were obliged to take our cows to the barn and cut their feed and draw it to them there. I had read a number of articles in my agricultural papers, on the subject of soiling, and, as an alternative, I accepted that mode of feeding, though I must acknowledge that it was with grave misgivings. It seemed to me that the amount of labor necessary to cut and bring to the cows their daily food, would require at least an extra hand. As I think of it now, it seemed like labor thrown away. That was before I had given it a trial. But now I am heartily glad, for it has saved me at least three thousand dollars, which I should have had to expend in fencing, had I not been compelled to change my course. For, since my first attempt at soiling, I have not built a rod of fence, and am taking down that which I had built. As I have before mentioned, my farm contains two hundred and forty acres. I have now but three fields that are fenced, and I intend removing some of them another year.

The plan I have adopted in growing and feeding soiling crops, may be stated as follows: Beginning in the fall, sow two or three sowings of rye,—enough to last until clover and grass is fit to cut in June. I find that an acre of rye or oats will feed my eighteen cows little more than a week. I have also three sowings of oats, two acres each time of sowing, two weeks apart. Following the oats, I have eighty acres of sweet corn, planted at different times, which I use for soiling, after plucking the ears for a canning factory, at eight dollars per ton, putting the surplus in shocks for winter feeding. We have in our barnyard a large feed rack which is filled every morning, fresh from the field, the cows running to it at will in the yard. The extra labor required to cut and feed the entire eighteen head, I have found to take from 1½ to 2 hours per day. The cutting is done with a scythe or corn cutter, as the case requires, by a man, with a boy to assist in loading and driving the team.

From May 1st until October 1st (twenty-two weeks) my entire herd has been kept from what has grown on seventeen acres. I feel safe to say that forty acres of pasture would not have sustained them in as good condition and flow of milk. My cows are all healthy and, considering that they are all milch cows, are in excellent condition.

As to soiling sheep, my experience is limited to one season's trial, which was, however, very satisfactory,—so much so that I am inclined to adopt it altogether as soon as I can get my farm in proper shape. I had, at the time referred to (summer of 1880), 90 ewes and 75 or 80 lambs. They were turned in the field to pasture during the fore part of the season, where it was intended to plant corn. On removing them to another field poorly fenced, they became so troublesome, and at that busy season of the year when there is no time to devote to rebuilding fences, I finally had them shut up in the yard adjoining their winter quarters, intending only to keep them there for a short time, until we could rebuild or repair the fence. We were at that time feeding our horses from grass mown daily from the orchard near by, conveying it to the barn in a wheelbarrow. In the same way, the sheep were fed in their winter feeding racks, allowing them the run of the yard referred to, two barrow loads a day, morning and night. Three pecks of poor beans and one peck of corn, mixed, was fed at noon. They kept in fine condition. In fact, I never had sheep do better. I was keeping them for a mere nothing. Sold all the lambs and forty ewes to the butcher. Could not say how much land was cut over per day to feed them, as the farm teams were fed from the same cutting.

Now, sir, I have given as you requested, my experience in soiling cattle and sheep. I should not consent to have you publish my doings, as I am but a beginner myself, were it not that I entertain the hope that it may lead and encourage others to try. If what I have said shall accomplish that much, I shall feel amply paid for my effort.

Yours, Respectfully,
C. H. FRANCIS.

I had the pleasure of visiting Mr. Francis' farm early in June last and suggested to him that he might like it better and find it more economical to feed his cows in their stalls where they would waste less and could be protected from the sun and the flies. Some time after my return, I received the following letter :

Middleport, Oct. 29th, 1881.

F. S. Peer,

Dear Sir ;—A few days after you were here I commenced feeding the cows in their stalls. I am fully satisfied that the two acres of oats, referred to in my letter on soiling cattle, would have fed the eighteen head three weeks in their stalls. I am fully satisfied that fourteen acres of rye or oats would keep the herd the twenty-two weeks, and, such a season as this has been, they would not do as well on fifty acres of pasture. I shall remove the rack in the yard and feed all in the stalls. I am surprised at the amount of feed that was wasted in feeding in the rack.

Respectfully Yours,
C. H. FRANCIS.

WINTER SOILING.

ENSILAGE.

"France, Germany, and other portions of Europe have practised summer soiling for centuries." For winter feeding, however, they have been obliged to cure sufficient fodder to keep their stock through the winter. There were numerous objections to this kind of food. The same grasses on which a cow would give an abundance of milk, from which rich, golden butter, of a very superior quality was made during the summer, if taken from the same field, if you please, cured and fed to her in the winter, produced a far smaller quantity of milk, and butter without color or flavor. Again, their young cattle would thrive and grow rapidly during the summer, while the same feed, cured for winter use, barely kept them from going back, to say nothing of improvement. Indeed, the farmer feels quite satisfied if his stock "hold their own" during the winter, even with the assistance of grains. If it were possible to supply stock in winter with such succulent and nutritions food as they obtain during the summer, on which they thrive and even fatten, the difficulties referred to would, in a great measure, be overcome. The advantages derived from securing a crop of grass or forage when in its most succulent state, and thus preserving it for winter consumption, are

1st.—That the farmer is enabled to continue the soiling system throughout the entire year.

2d.—More stock can be kept on the same number of acres, or the same on a less number.

WINTER SOILING. 115

3d.—*The accumulation of manure.*

4th.—The greater production of milk, butter, beef, wool or mutton.

5th—*The more thriving condition of young stock.*

6th.—Less money must be invested in barns, in which to store winter food.

With all these advantages to be gained, it is no wonder that Mother Necessity gave birth to an invention that would overcome nearly all obstacles and make attainable the benefits sought.

As we have already considered most of the advantages to be found by adopting the system of ensilage under similar heads on the subject of soiling, we will take only a passing notice of them, considering them only as far as their effects produce further proof of the practicability of conducting the soiling system throughout the entire year.

1st.—All animals prefer their food green. Analysis makes no account of the juice of plants; but calls it so many pounds of water. Why is it then, when it is fed with the water in it, that it produces far better results in the attainment of beef, milk, butter, mutton and wool, than when cured, which is simply the taking out of what the chemist calls water, by the process of evaporation?

Again, if none of the value of forage crops is lost by curing, it must be locked up in the woody fiber of the plants and cannot be assimilated until thoroughly cooked (steamed) or soaked. I am inclined to the latter opinion, but, nevertheless, the water, or juice, though it may possess no value of nourishment more than well or rainwater, yet, if it is the means of holding the properties of plant-food in the condition that renders them capable of being easily and entirely assimilated by the animal which consumes them, it should be credited with performing all that it is necessary for man to do by cooking, &c., in order to obtain the full benefits of it. And, even when we have been to the expense of steaming, soaking, &c., which no doubt places all cured feed in a better condition for the animal to extract its feeding value, yet no advocate of cooked food ever pretends to say that it is thus fully restored to the condition that produces the same results as when fed in its green state. In regard to the value of water as food for farm stock, Dr. M. Miles, (*American Agriculturist*, p. 53, 1882) says: " It will also be readily seen that the supply of water should be constant, or at least frequently repeated, to secure uniformity in the fluidity of the blood and the various secretions. Water must be recognized *as a food*, and it should be given with the same regularity as other food. Magendie found that dogs, supplied with water alone, lived from six to ten days longer than those that were deprived of both food and water; so that water has undoubtedly an important function to perform in the system, aside from the dilution of other nutritious substances. The addition of green food in some form, to the winter rations of our farm animals, will be found advantageous for many reasons, and the amount of water required by them as drink will by this means be diminished."

It always seemed to me, when walking over a field of newly mown hay, the air loaded with its delicate perfume, that there was a loss going on which I had no power to restrain. Then there is a great difficulty in securing the crop, (especially clover) waiting sometimes for better weather, the crops rapidly passing from its most nutritious condition to a riper one of less value, or when in the process of curing a heavy shower robs it of a large per cent. of its value, sometimes leaving it little better than so much straw. To all of this the farmer *must submit.* But, if the crop is to be ensilaged, the farmer is master of the situation,—no waiting until the dew is off, no need of stopping for a passing shower, so far as its being a damage to the feed is concerned. In fact, a heavy dew or a light shower, unless a hindrance to the work, is more to be thankful for than otherwise.

The cutting may be begun and completed when the forage is in its best and most perfect condition and, as we shall presently show, at a less expense than if cured as hay.

2d and 3d.—In regard to keeping more stock and the attainment of manure. All that might be said here on this subject, has been mentioned under " Summer Soiling."

4th.—In regard to the greater production of milk, butter, beef, wool and mutton, there can be but one opinion. If stock are provided with comfortable quarters, there is no reason why the manufacturing of butter may not be carried on as successfully and as profitably, if not more so, during the winter, as in the summer months. Good, fresh winter butter is worth from 10 to 15 cents per pound more than that of the same quality, made during the summer. The necessary labor may be had for from one-half to three-quarters less. It greatly reduces the profit on butter where farmers are obliged to hire their help, especially during the busy season, when the help in the field is worth $2.00 per day.

Again, there is no annoyance from flies, and all the unpleasant incidents to butter making, that occur in hot weather. There is also no necessity for ice, &c. My own plan since adopting the system of winter soiling (with 10 or 14 milch cows) is to have (as near as possible) one coming fresh in milk every month in the year. Thus I am able to supply my customers with butter, fresh every week or two, *uniform in quantity and quality*, during the entire year.

In regard to the production of beet, I have no personal experience, but from all reports it is highly recommended. For sheep, I think very highly of it, especially where it is desirable to have early lambs. The ewes giving more milk, and it being a food that young lambs will learn to eat, when but a few days old.

5th.—In regard to a more thriving condition of young stock when fed ensilage, in comparison with dry fodder, I am also very much pleased at the results. I never had young stock do as well as last winter, 1880–1 (which was my first year's experience) although confined to their winter quarters for over 9 months.

6th.—Less money invested in barn room in which to store winter feed.

A cubic foot of ensilage, if cut at the proper time, and well pressed, will weigh about 50 lbs. Therefore one ton would occupy only 40 cubic feet. A ton of hay in mow or stack, is estimated to occupy 525 cubic feet, or 13 times as much space as a ton of ensilage; but, as it requires about two tons of ensilage to equal in feeding value a ton of hay, the ensilage would require less than one-sixth as much space. In other words, if a farmer were to build for the storage of hay, or a building in which to ensilage the same, the former (of the same material) would cost about 6 times as much.

THE DISCOVERY.

The discovery of the method of preserving fodder green, is attributed to Monsieur Goffart, of France. After many experiments, and the expenditure of considerable money, his achievments were crowned with success and honor. For years he held to the idea that the fodder should be at least partly cured, and that it should be put in the building in alternate layers with straw, until, more by chance than otherwise, he discovered that the curing process and the use of straw, were the very causes, more than anything else, that held back the discovery, and that perfect preservation was only attained when the forage was brought directly from the field as fast as cut, passing it through a feed cutter, which, making it finer, admitted of closer packing.

The principle is to exclude the air. The forage should, therefore, not only be very closely packed, but all the cells containing the juice should be kept full. These cells or pores are emptied by evaporation or curing, and air takes the place. The ordinary pressure is not sufficient to expel it. The plant, therefore, contains within itself sufficient air (oxygen) to destroy it.

THE SILO.

"Silo" is a French word and means "a pit" (*en silo*, in a pit), and this is the origin of the word, "ensilage," the manner of storing green fodder in a pit, or "in an air-tight manner." The material used in this country at present is generally masonry, or concrete with cement floors, and walls plastered with the same or water-lime. The inner surface of the walls should be perpendicular and smooth, so that they will offer no resistance to the settling of the forage. Some report very satisfactory results from silos built entirely of wood (matched boards or planks), others, where the soil is such as to exclude water (rock or clay) preserved forage successfully by digging pits or trenches, afterwards covering with earth.

HOW LARGE TO BUILD.

In estimating the size of the building necessary to supply your herd and flock, 1 head (1000 lbs.) will consume in a day 2 to 2½ cubic feet of ensilage. This is a full ration. If other feed, such as hay or coarse fodder, is to constitute part, it may be taken into consideration. It is also well enough to have room so that the number of your stock may be enlarged. At 2½ cubic feet per day, 1 head will consume in 6 months (180 days), which is about the length of our winters in this section) 450 cubic feet, to which add, say 50 cubic feet for settling, or 500 cubic feet. Multiplying this by the number of head, will give the size of the building required, in cubic feet.

WHERE AND HOW TO BUILD.

The silo, if possible, should be so built that a door from it will open directly into the cow-stable, and the bottom, or floor of the silo should not be more than one or two steps below that of the stable. They may be built either above or below ground, or partly above and partly below. The position of the stables should determine. It is all useless to dig, as some have, ten to fifteen feet below the surface, making it necessary to lift 200 or 300 tons up and out. Others think that, as it is necessary to fill the building by attaching a carrier to the feed cutter, and elevating it over the top of the wall, it must necessarily be taken out over this wall, and they are obliged to rig derricks or be to great task to get the feed out.

Now this is a mistaken idea. It is all right to fill the silo by elevating the cut forage over the top of the wall, but it may be taken out through a door anywhere in the wall where most convenient and with the least labor. In which case, the door must of course be sealed air-tight, which I have found no trouble in doing, by simply closing the door, which is hung on the outside of the casings (walls 20 inches thick), boarding up on the inside of the door casings with matched boards, even with the plastered surface of the wall, thus leaving a space the thickness of the wall, which I fill with sawdust, packing it tight; nothing more. The bottom of the silo is about 18 inches below the top of the ground. The walls on the inside are 15 feet high, and one end of the building stands against a side hill, where the top of the ground is about 8 feet above the bottom of the silo. The filling is done at this end by attaching a 12-foot carrier to the cutter, the cutter standing on a raised platform 3 feet above the ground. If the silo is not near the barn, or if desirous to draw the ensilage to a distance, the floor should be on a level with the ground outside, and the door wide enough to back a wagon or cart inside and fill it. I know of a silo 30 or 40 feet from the stables, six feet under ground and six above, where the ensilage is all elevated over the top of the wall, and carried, a bushel at a time, to the stock. I repeat, it is a mistaken and useless practice, that may be easily avoided by having the door through which the feed is to be taken out on a level, or nearly so, with the stables, and if possible, open directly into them. The deeper the silo, the better, as it costs no more to roof it, and the planks and weights required to press five feet deep, would answer as well for fifteen or twenty.

My own silo was formerly an old stone carriage house. When I became convinced that ensilage was not a humbug, that by its adoption my stock would be supplied with green food during the entire year, and that it would be to a great degree a continuation of the soiling system, (the advantages of which I was already familiar with), I was not long in making up my mind, and therefore needed little, if any, encouragement to undertake it, and soon set to work remodeling the old barn, taking out the hay loft, floor and stalls below, walling up the doors and windows, except the door already referred to. It has a storage capacity when filled, sufficient for 25 head of full-grown stock for six months.

NUMBER OF TONS PER ACRE.

Last season we estimated by weight of a cubic foot (50 lbs.) that we had 160 tons of ensilage which grew on 5½ acres. It was the greatest growth of fodder I ever saw. The seed was the Western Dent variety, growing from 7 to 9 feet high. Some claim as high as 40 to 60 tons per acre, but I must say it was a mystery to me and many others how an acre of land could produce twice the amount that was grown upon the 5½ acres referred to.

CROPS FOR ENSILAGE.

Any green forage may be ensilaged, if cut at the proper time, *i. e.*; when in blossom, or several kinds mixed, providing they flower about the same time, but corn fodder, on account of its exuberant growth, is usually preferred to other grasses. It will yield from 3 to 5 times as much feed per acre, and is not particularly exhausting to the soil, as shown by the analysis in respect to its nitrogen, phosphoric acid and potash, as compared with other green crops. I have grown corn fodder year after year on the same ground with equally good results.

Says Goffart, (Brown's translation, p. 15), "Some of my finest maize occupies a field which, during the past eighteen years, has borne fourteen harvests of that plant, without giving any signs of weariness; on the contrary, the latter yield is better than the former."

Some advocate following corn with rye in the

fall, ensilaging the rye in the spring for summer feeding, and sowing the ground again to corn, thus securing two crops from the same land in one year. This may easily be accomplished, so far as the rotation is concerned, but will doubtless require heavy manuring. A better plan would be to sow the field to rye after the corn is cut in the fall, and plow it under as a manure, the following spring in time for corn. If analysis is worth anything, the rye would supply the field with nearly enough plant food to grow the corn.

METHODS OF CULTURE.

The corn should be sown in drills from 2 to 3 feet apart. If desirable to fertilize with barn-yard manure in the hill, the rows should be furrowed out with a shovel plow, the manure sprinkled in the track and covered lightly with soil before putting in the seed. Or an easier and less laborious way of attaining the same results, is to top-dress after the corn is up. On some soils it may be advisable to plow the manure under, especially if very coarse; but experience has taught me, that upon my farm, top-dressing produces far better results. The corn may be put in with a common (nine tooth) field grain drill, by altering the tubes or spouts that conduct the grain from the hopper to the drill teeth, as follows: closing No. 1, and letting 2 and 4 discharge into 3; closing No. 5, and letting 6 and 8 discharge into 7. If the drills are 8 inches apart, this will make the rows of corn 32 inches apart, putting in two rows at a time, the wheels serving as a guide on return bouts. The ground should be rolled. As soon as the corn begins to prick through the soil sufficiently to show the rows, it may have a dressing of plaster (gypsum) with good results. If composted with ashes, fine manure and muck, so much the better. In selecting a variety of corn for ensilage fodder, the kind that produces the most leaves should have the preference. Principal among the varieties used for this purpose, may be mentioned the Southern White, Western Dent, and Blunt's Prolific.

Last season I sowed at the rate of 2, 2½ and 3 bushels per acre, in drills 32 inches apart. I am satisfied that 2 bushels per acre is sufficient. The corn should be cultivated with the horse-hoe until the ground is completely shaded. If the soil is not stony, the cultivating may be done until the corn is 4 or 5 inches high, with a smoothing harrow, such as the "Thomas," manufactured at Geneva, N. Y. Corn, either for soiling or ensilage, should never be sown broad-cast, for reasons already given.

CUTTING AND STORING.

The cutting may be done with a stoutly built self-raking reaper, which has already been referred to,—two men to bind the fodder, assisted occasionally by the one who reaps. Two men, each with a one-horse lumber wagon, draw the fodder to the silo alternately, so that while one is loading at the field, the other is unloading at the platform, where stands a powerful feed cutter, through which the fodder is passed, cutting it ⅜ of an inch in length, as fast as two men could feed it. The carrier (12 feet) attachment elevated it into the silo, where one man is employed spreading and treading it down. Some recommend using a horse or mule for this purpose, but think it is unnecessary.

Dr. Baily draws the fodder from the field in dump carts without binding, unloading at the silo by dumping. Others have ropes underneath the load, to which a horse is attached and the whole load pulled off at once. My plan for the coming season, is to load on to a flat rack with stakes before and behind, without binding, using 12 or more slings (ropes) which, with the aid of a derrick or horse could be lifted upon the table of the feed cutter with ease. It would save at least the labor of two men and much heavy lifting.

PRESSING.

After the ensilage is all in, or the building filled, there is generally spread evenly over the surface, from 1 to 2 feet of uncut straw, after which a floor of 1½ or 2 inch plank (rough) is laid upon the straw. It was formerly thought that it was necessary to have the planks matched, the object being to press the ensilage so hard and close that there is no room for air, more than to attempt to exclude it by having an air-tight cover. This plank covering should be weighted with at least 100 lbs. to the square foot, which may be obtained by using stone, bags of sand, or earth. Dr. Baily saws kerosene barrels through the middle, fastening to each half an iron bail. He fills the tubs with sand or stones and, by means of a derrick, places them on the plank. The same derrick is used to unload when the ensilage is required for feeding. Others use screws, some from above the ensilage, others by means of inch iron rods passing up through the ensilage and through a timber passing lengthwise of the building and on top of the plank, by means of a nut on the end of rods, which is tightened three times a day, for two or three weeks.

Care should be taken not to press too hard, or the juice of the fodder may be extracted. For some reasons the weighting process is the best. When once applied, it needs no further attention; its pressure is continued day and night; it holds all it gains, and there is no danger of pressing too hard. On the whole, I am inclined to think Dr. Baily's plan the best and most economical.

OPENING THE SILO.

Six or eight weeks should elapse after filling, before the silo is opened, to allow sufficient time for the ensilage to become thoroughly packed. The sealed door referred to may now be opened, taking out the filling and boards on the inside, using only the outer door, taking up two or more plank the whole width of the silo. With a hay knife cut down, leaving a perpendicular wall of ensilage. This cutting should extend to the floor as it may be needed to feed. There will be no difficulty arising from leaving the ensilage exposed, providing another cutting follows within a week or two.

FEEDING.

It is generally considered that two tons of ensilage are worth one ton of hay. Every kind of farm stock prefer ensilage, to the best kind of dry fodder. This has been repeatedly proven by placing hay and ensilage side by side in the yard, leaving the cows to choose. This is perhaps a stronger argument in favor of its superiority over dry fodder, than those advanced by some skeptical farmers opposed to ensilage, who have had no experience with it. I have no practical knowledge of chemistry, therefore will not attempt to dispute the point that there may be found as much feeding value in the same amount of forage ensilaged or cured, that is to say, of two acres of clover of equal growth, ensilaging one and making hay of the other. Nevertheless, I do maintain that the animal food in the ensilage will be more easily assimilated by the animal, and thus be more profitable to the feeder. Analysis also shows that some fields contain an abundance of plant food, and yet do not produce as great a growth of plants as a neighboring field that has less plant food. Why? Simply because in the former case, the plant food is insoluble, and the plant fails to assimilate sufficient food to produce as good results as in the second case, where there is less plant food, but more soluble, and therefore more easily assimilated with the nature of the plant.

The greatest and most characteristic benefits to be derived from ensilaging green fodder is, that by the use of corn, and other grasses, that produce great quantities of feed, there may be from 6 to 10 times as much feed obtained from the same number of acres. Ground that will pro-

WINTER SOILING.

duce one ton of cured hay per acre will, I think, produce 12 to 20 tons of ensilage, equal to 6 or 10 tons of hay.

On this question, *i. e., the saving of land,* it seems to me there can be no hair-splitting discussion, but that there is a profit, *clear, distinct and undeniable.* It may be asked if it would not be cheaper to cure the corn fodder in the field by shocking it, than to cut and put it into a silo. I answer emphatically *no,* and for the following reasons: It costs no more to cut and bind it, than to cut and shock it. It must in either case, be drawn to the barn. It is certainly preferable to secure it all at once, where it is in easy reach of the cows, where the feeding may be done with ease and comfort, let the weather be what it may, rather than to leave it standing in the field, or in small stacks near the barn until wanted for use, exposing it to the weather, freezing fast to the ground, often being blown over, and covered with snow and ice when wanted to feed.

Any farmer that has had one year's practical experience in this latitude, trying to cure corn fodder for winter use, is not apt to repeat the experiment, and this is not all, much of the value of the food is thus wasted, or in the curing process, and also becomes less valuable for food. I have tried both ways, and can unhesitatingly say that the system of ensilaging fodder is more economical than *trying* to cure it.

Before opening the silo, we had fed, morning and night for two weeks, all the cornstalks the cows would eat, and roots and wheat chaff at noon; but our butter was white and lacked flavor. It was a poor substitute for a first-class article.

We opened the silo November 12th, and forthwith began to feed ensilage twice a day, morning and night, and straw and roots at noon as before. At the fourth day's feeding, the quantity of milk was very nearly doubled, and as to the butter, it was equal in flavor to, and in color only a few shades lighter than that made in summer from green food.

We continued feeding in the last described manner for five or six weeks, and with the same pleasing results. Then, in order to dispose of our coarse fodder, we substituted cornstalks and barley straw (cut and fed dry) for the morning feeding of ensilage. The decrease in the amount of milk was very marked, the yield shrinking about one-fourth. The color was also considerably lighter. Hoping to make up for this deficiency, we added two quarts of corn meal per head to the ration of dry fodder, but it did not fully compensate for the full feeding of ensilage. The quality and color of the butter was not equal to that made from ensilage and roots alone, the quantity however, was increased to about the same. The following shows the cost of ensilaging five and one-half acres of corn fodder, or one hundred and sixty tons:

	Total,	Per acre	Per ton.
7 men. $1. per day, 4 days,	$28 00		
Boarding 5 men 2 meals per day,	6 00		
Engine and Engineer, $4.00 per day,	16 00		
Fuel and Oil,	4 00		
Total cost of labor to secure 160 tons,	$54 00	$ 9 82	33
Cost of seed, fitting the ground and cultivating,	27 50	5 00	17
Total Cost,	$81 50	$14 82	50

WINTER SOILING.

Upon 5 acres, in a field of 12 acres, I grew 30 tons of ensilage per acre. The remaining 7 acres, cut for hay, produced less than one ton of hay per acre. This was not a full yield, as the clover grubs were very numerous. However, the field has never, under the most favorable circumstances, produced over 1½ tons of hay per acre. Therefore I feel safe in saying that land capable of producing 2 tons of hay per acre, will grow (without manure) 30 tons of corn fodder per acre, if the proper kind of seed is sown.

I have had no way of personally testing the feeding value of ensilage as compared with hay. As before stated, those that have written upon the subject, claim that they have discovered by actual test that "2 tons of ensilage are equal to 1 ton of the best kind of hay." But simply for the sake of comparison, and not to insinuate that I doubt the truthfulness of the assertion, let us suppose that it will require *four* tons of ensilage to equal one of hay, and as a basis of calculation, say that an acre of ground capable of producing 30 tons of ensilage per acre, will produce 2 tons of hay. The comparative value of the two crops will be found in the following table.

Table showing the value of 1 acre of Corn Fodder for Ensilage, as compared with 1 acre of Grass for Hay.

	Ensilage Dr.	Ensilage Cr.	Hay Dr.	Hay Cr.
2 tons of hay, feeding value at $15. per ton,				$30 00
Seed 1 acre,			$ 1 00	
Cost of curing and delivering to barn,			$ 2 50	
30 tons of Ensilage, equal to 7½ tons of hay at $15. per ton,		$102 50		
Seed, fitting the ground and cultivating,	$ 5 00			
Labor to cut and secure 30 tons of Ensilage,	$ 9 82			
Total,	$14 82	$102 50	$ 3 50	$30 00
Balance,		$ 14 82		3 50
Net Value,		$ 87 68		$26 50
Balance,		26 50		
Profit of 1 acre of Ensilage over 1 of Hay,		$ 61 18		

The amount of feed that may be produced from one acre by growing upon it a crop of corn fodder instead of hay, is where the farmer may look for the great profit of ensilage over dry feed, and this profit is so distinct, that he need not hesitate to adopt the system on account of the fact that analysis shows that there is no more value in an acre of grass ensilaged, than an acre of grass cured for hay. As I have before stated, I have no reason to enter into a hair-splitting discussion upon this view of the subject, because, by growing corn fodder and securing it, I am able to show a profit by the system of ensilage that defies all contradiction, and is clear and undeniable.

The illustrations on pages 134–5, are designed to represent the ground plan and elevation of a moderate sized barn,—upright part 30x40 feet ; shed 30x35 feet ; manure shed, 14x30 feet.

The plan of cow's stable is the same as that represented on a previous page, where the measurements are given. P is the feed alley ; M M are mangers ; d d are the drops; P P passages behind the cows, which open by rolling doors into the manure shed ; C is the cart in which to draw ensilage from the silo along down the feeding alley P, and in front of the mangers of the box stalls B B, and the feeding trough of the sheep shed, making it convenient and easy to feed all the stock.

When soiling, the loads of feed are driven on to the barn floor above and thrown down through the floor into the feed alley P, at C. W is a well or spring on the line of a fence separating the yards.

SOILING.

WINTER SOILING.

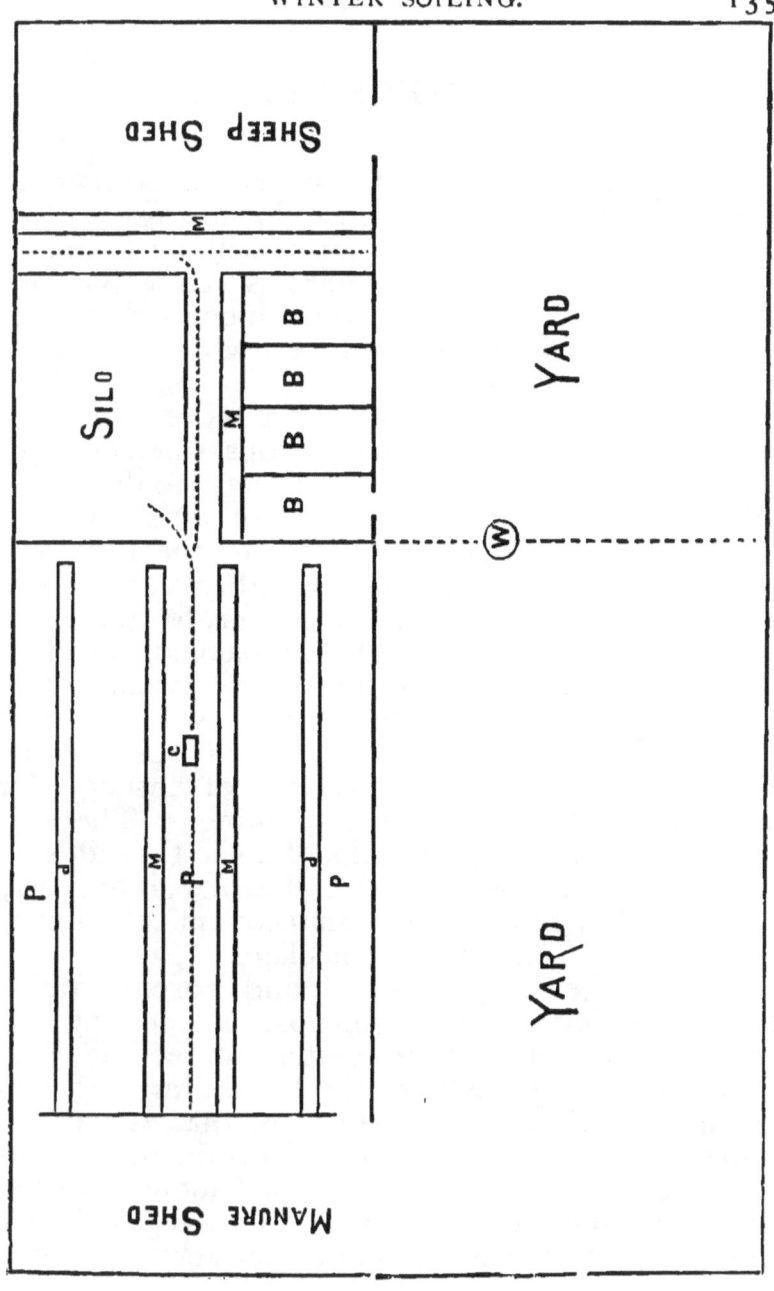

ENSILAGE SOILING.

In regard to adopting the system of ensilage as a method of summer soiling, I do not agree with some writers, who claim that it will be found as economical as soiling. There is really but one thing in its favor, whereas the objections are numerous. The only gain is, that a whole field may be secured at one time.

The objections are, that it is more expensive than soiling; when soiling, the crops, when brought to the barn, may be delivered directly to the stock, and the labor may be performed by a boy, whereas, if ensilaged, besides drawing to the barn and feeding, it must be passed through a cutting machine, requiring some kind of power, to run it. It must also be spread, and tightly packed, covered, weighted, uncovered, cut down, and taken out of the silo, without adding to the quality of the food.

This extra labor makes it, in my opinion, impracticable to adopt the system of ensilage for summer feeding, instead of soiling. There is nothing within the four walls of a silo that adds to, or destroys the feeding value of forage, if properly secured; but simply from an economical point of view, soiling is superior to ensilage.

However, ensilage will be found more profitable than pasturing. For the purpose of showing the relative value of the three systems of feeding, *i. e.*; by pasturing, by feeding and by soiling, we will estimate, as a basis of calculation, that it will require 2 acres of land capable of producing 2 tons of hay per acre, to support 1 cow 6 months, or 180 days, and that land equally productive, will produce 30 tons of ensilage fodder per acre, or that it

ENSILAGE SOILING.

will require ⅛ of a square rod of corn fodder (as previously shown) to support 1 cow a day (or 24 hours). From this hypothesis, we may easily find the comparative feeding values in dollars and cents, of 2 acres of land under each of the three systems of feeding.

Table showing the Comparative feeding value of Pasturing, Ensilage, and Soiling.

	Pasture.		Ensilage.		Soiling.	
	Dr.	Cr.	Dr.	Cr.	Dr.	Cr.
Feeding value of 2 acres by pasture, 50 cents per week, 180 days,		$12 00				
Feeding value of 2 acres by ensilage, 50 cents per week, 600 days.				$43 30		
Feeding value of 2 acres by soiling, 50 cents per week, 640 days.						$45 50
Cost of seed for 2 acres pasture,	$2 00					
Cost of ensilaging 2 acres corn—60 tons at 50 cents per ton,			$30 00			
Cost of soiling 1 cow 640 days, at 1½ cents per day,					$9 80	
Cost of fitting the land and seed (soiling) $5.00 per acre,					$10 00	
Total,	$2 00	$12 00	$30 00	$13 30	$19 80	$45 50
Balance,		2 00		30 00		19 80
Comparative profits from 2 acres,		$10 00		$13 30		$25 70

The above table not only shows that soiling is much more economical than ensilaging forage for summer feeding, but it also serves to show the profit of soiling and ensilage over pasturing.

The table, however, does not represent the entire profit of the system of soiling and ensilage over pasturing, as the two former should be credited with

1st.—What is saved or gained by the increased quantity and quality of manure.

2d.—By the greater production of beef, milk or butter.

3d.—By the better condition and greater comfort of farm stock.

4th.—By the saving of fences.

And again, that most valuable consideration resulting from the increased quantity and quality of manure, *i. e.; the increased fertility of the soil.*

SYSTEM.

There is one thing especially necessary in conducting the soiling system *successfully*. It is not capital, as some might suppose, for men without capital are usually the first to adopt it. It is also unnecessary that a man should have a large farm stocked and equipped, because the system is equally well adapted to a limited number of acres.

It appears in a work recently printed by Orange Judd & Co., N. Y. City, on "Keeping One Cow," containing the experience of fifteen different owners of one cow each, all living on small places, usually village or city lots, that the greatest amount of land required to keep a cow one year, was 2 acres; the lowest, 1⅓ acres. The greatest amount required to keep one cow through the summer, from May 1st to Nov. 1st, was 1 acre; the lowest, was ½ of an acre.

Nor will those only be successful who live near large cities, where land is high. Whatever may be the condition of the land, it is safe to say that the amount of land that will keep one head by pasturing, will keep four by soiling. The rule works as well on cheap land, as on high-priced land, the latter not being necessarily more productive than the former. Therefore, if from land worth $25.00 per acre, a farmer sells as many dollars' worth of produce, as on land near the city, worth $200.00 per acre, the soiling system is as profitable to one as to the other. The difference in the profit from soiling, will be found in the productiveness of the soil, and not necessarily in the price of the land. If, on a farm worth $100.00 per acre, a farmer can keep 1 cow 1 year from an acre of land, and another, whose farm, on account of its location, is worth $200.00 per acre, but is only capable of keeping 1 cow a year upon 2 acres, the profit in soiling is greatly in favor of the farmer with the cheaper land, so far as keeping cows is concerned.

I mention this, because it is often stated, that "it may pay to soil, where the land is high-priced," and to show that the price of the land is not a sure indication that soiling will be found successful in proportion to its cash value. I can imagine, however, a farmer, under the most favorable circumstances, failing to obtain satisfactory results from soiling, for the want of *system*.

Without system, I can easily imagine that a farmer may soon become disheartened, and pronounce the whole thing impracticable.. For instance, by omitting to sow at the proper time, or the proper amount. Sowing too much at a time, the stock are unable to consume it in its most succulent state, continuing to feed until it becomes tough, when it is only eaten to satisfy intense hunger. By having too little, his cows must be turned into the field until the next crop is in condition, thus causing him to become dissatisfied.

Again, I can imagine a man with plenty of feed, putting, at one feeding, sufficient before his cows to last them all day: they breathe upon it a few hours, and nothing short of severe hunger will induce them to take it, in which case his stock would shrink in the flow of milk, and increase on turning them to pasture, which would lead him to say that the cows did better at pasture, and thus condemn the system.

Again, by not having properly constructed stables or stalls, they might become very filthy and unhealthy, and the cow would " long for pleasant fields and pure air," and this might lead the farmer to abandon the system.

Again, his manner of cutting and feeding might require more labor than the advocates of the system profess, and he might thus think that the system might be well enough for a farmer with plenty of capital, a "fancy farmer," a "book farmer,,'" but not for him.

Again, by his undertaking too much at once and getting everything all mixed up, I can imagine the last state of that man, as worse than the first.

But by so systematizing the work, that every want will be supplied, I can assure, yes, guarantee, any man success. He need not, necessarily, follow in detail the plan I have laid out in the previous pages, for it is not so perfect but that it may be improved. I know, if closely followed, the system will lead to success, therefore I may be pardoned for saying, that until you can learn by *actual experience*, a better way, I would advise the beginner to adhere to the plan I have pointed out, in all essential points. I have tried many things that thought has suggested, that indeed looked as if they would result in improvements, but, when put to the test, were found wanting. I have no fear of contradiction from those who have successfully practiced soiling, when I say that the principal requisite necessary to success by soiling, is *system*.

The work of sowing, cutting and feeding, should all be placed in the charge of one person, who can be relied upon to do the work as directed, and when the daily routine is once established, it will be found much less laborious than it seems to be. The labor is comparatively light; it may be performed by a boy; but nothing can be left to chance.

When the proper time comes for sowing, the

work must be done. The cutting must also be attended to when the crop is ready. The feeding also must be regular and uniform in quantity.

The stables cannot be neglected for a day or two without cleaning. It is unnecessary to say what the results of this neglecting an easy task would be. With a little practice, and by a person not entirely destitute of ability to work systematically, he cannot easily fail of conducting the soiling system with profit, and also to enjoy the many advantages that it affords. I have never heard of a man who once thoroughly adopted the system, who was not, ever afterwards, decidedly pronounced in its favor.

The methods employed, have a tendency to lead the farmer to the performance of all other labor on the farm with more system, and to conduct the profession more on business principles, and demonstrate more clearly the force of the old adage, " *Time is money.*" He is not a husbandman in the true sense of the term, who fails to comprehend that, by a *systematic* husbandry of time, he is as truly practising one of the highest arts of his profession, as when plowing, sowing or reaping.

I do not mean that husbandry of time consists in constant hard labor during every hour of the day, nor in obtaining from farm help an extra hour's work before sunrise or after sundown, but that it does consist in so systematizing the work, that his help will accomplish as much in 10 or 12 hours, as in 14 or 16.

The farmer does not accumulate wealth by grand speculations and by rapid strides, but by turning to account the minor details of his business.

Therefore, of all men, the farmer should conduct his affairs with the most rigid system. I know of no profession that suffers more for the want of it. It is due to this, more than to any other cause, that farming, as an occupation, is held in disrepute by certain classes of business men of other professions, and justly too ; for if any other business should be conducted in the same desultory manner that usually characterizes farming, failure would inevitably ensue. On the contrary, were the farmers to conduct their business with as much system as is found in every other pursuit, it would not only be more remunerative, but possess a greater attraction for young men with business faculties, who now forsake the farm, and what they term its drudgery and humdrum, to engage in a pursuit that offers more attractions, greatly on account of the systematic way in which it is conducted.

Thus they often throw away the "*pearl of independence,*" for a life of unremitting toil, harassing and perplexing, in comparison of which labor on the farm is a pleasure.

EDUCATION FOR FARMERS.

As Mr. Stewart says, in conducting the soiling system successfully, "the need is more for head than hand work."

I believe he might have extended the remark to every branch of agriculture, especially where the price of land is necessarily high. The day has gone by, in the older States when a man can follow farming because he does not know enough to do anything else. It may be done in the west, where land may be had for the asking, and so productive that by "the slightest effort it will produce an abundant harvest," but in the east, it is not only essential that a farmer should possess a knowledge of how to produce a crop from the soil, but how to leave the soil in as good condition as before the crop was taken, or better. This, in my opinion, is good farming; while he who harvests a crop at the expense of the soil, is not a true *husbandman*.

Farming is an honorable profession, but he who tries to obtain by it, something for nothing, is never a credit to his profession. There seems to be among some classes of farmers, a great antipathy to what they term *book farmers*. Why may every other man learn what pertains to the advancement of his business from books, and not the farmer? We point with pride to this or that man in the medical profession, and say he is a well read physician; to a lawyer, and say he is a well read attorney; to a citizen, and remark that he is the best read man in the place. These are chosen and preferred for their learning, and their excellence is measured by the number of books they have mastered.

Again, why should farmers subscribe for two or more papers devoted to *politics*, religion or science, and read them diligently; papers devoted to every subject but one? Why purchase books of fiction, books pertaining to all subjects but one, and that one *his own business?* Why does he consult his neighbor as to his methods of growing a certain crop, and follow his example, when, if the neighbor should write out his experience in book form, it would be denounced as book farming? Whence do farmers' sons get the idea, that as soon as they obtain an education, there is no use for it on the farm? They are sent to school; taught chemistry, botany, engineering and surveying, but, from their fathers' examples, they have learned to think that such an education may do well enough for a book-keeper, or a dry goods clerk, but to apply such knowledge to an agricultural pursuit, is all wrong; 'tis book farming, and yet it is knowledge that can be put to practical use only on a farm.

Do farmers mean to acknowledge that their profession is less noble and intelligent than others?

What is there in farming that requires a man to be ignorant? Must a farmer, in getting on in the world, move backward like a crab, or, as Mark Twain says of the inhabitants of the Azores Islands, among whom all efforts to introduce new and approved methods of farming have failed, "The peasants crossed themselves, and prayed to God to shield them from all blasphemous desire to know more than their fathers did before them"?

These questions I will leave the reader to solve. However, I will venture to suggest as a remedy, a better education for the future farmer.

The great problem of feeding and clothing the millions depends upon the success of agriculture, and requires of its followers, a knowledge that embraces a wider and more liberal education than any other pursuit.

Said the late President Garfield, "At the head of all the sciences and arts; at the head of civilization and progress, stands, not militarism, the science that kills; not commerce, the art that accumulates wealth; but *agriculture*, the mother of all industry, and the maintener of human life."

CONCLUSION.

As I look back to the time when I did not practice soiling, and reflect upon the condition of my farm then and now, I am greatly encouraged to believe, that at no distant day, I shall see the fertility of the old farm restored, and again witness its fields bountiful in harvest, and its flocks and herds greatly increased. Judging from the past, the time will soon come when, by soiling summer and winter, I shall be able to keep equal to 100 head of stock upon 50 acres of land, during the entire year, and devote the other fifty to growing grain.

During the few years that I have practiced soiling, I have been enabled to treble the number of farm stock, where, by pasturing, 12 head was a greater number than the farm could profitably support. I have also been enabled to double the number of acres devoted to crops. This was accomplished by soiling four months during the summer, and preserving five and a half acres of ensilage for winter feeding. I have never had a sick cow or horse since I began the system of soiling, and have never lost a sheep or lamb while soiling them, nor had one sick, from any cause attributable to that method of feeding.

Thus the result, in every respect, has not only been most satisfactory, but I think no consideration would induce me to again pasture my stock during the whole season. Nor is my experience inconsistent with the statements made by European and American writers on the subject.

I could have no object in presenting this subject in other than its true light.

To others who may begin farming, or who may now be occupying farms similar to my own, or to any one who has faith in barnyard manure, as the best means of enriching the soil, and would be glad to obtain it "at a cheap and easy rate"; to the farmer contemplating building new fences or repairing old ones; to the farmer who is about to buy more land; to the possessor of an acre or two of land, living near the city or village, who would be very glad if he could keep a cow; to all, I should like to say, give the system of soiling *a thorough trial*, and if it does as well by them as it has by me and a great many others, I shall feel that my efforts in trying to persuade them to adopt the system, will merit at least their good will and approval.

And to the young men, particularly farmers' sons, to whom this book is dedicated, I wish to say, that there is no calling that promises to its followers the same pleasing, honorable, and independent life, as *intelligent* farming.

Though in some respects unpleasant, and means sometimes dirty work, dirty clothes, coarse boots, hard, black hands and bronzed faces, and white gloves and rose water perfume will not disguise the fact that you are "from the rural districts," nevertheless you have a guarantee to better health, better opportunities for mental improvement, for more generous hospitality and social intercourse, than in any other livelihood.

Above all, be intelligent in your business, and your situation, if even a humble one, will challenge the respect of all whose opinion is worth having.

The day has gone by, if it ever was in this country, when a young man need hang his head because of the humbleness of his vocation, if it be a useful one. Do not leave the farm for a profession or clerkship in town, unless you are specially adapted by natural tastes and inclinations to follow it. Do not throw away the pearl of independence for toil unremitting, harrassing and perplexing, in comparison to which, work on the farm is a pleasure.

Thousands of men die of broken hearts, who would have lived happily at the plow.

Thousands more, " look upon the healthful and independent calling of a farmer with chagrin."

"Gen. A. H. S. Dearborn, of Boston, who had long been acquainted with the business men of that city, gave it as his opinion, that only three men, out of every hundred doing business, were successful. * * * * A person who looked through the Probate Office of the same city, found that ninety per cent. of all estates settled there were insolvent. Yet more discouraging to the young man who would enter commercial business, were the conclusions of Governor Briggs and Secretary Calhoun, who a few years ago gave it as their deliberate opinion, after diligent inquiry, that out of every hundred young men who came from the country to seek their fortunes in the city, ninety-nine failed of success." ("Getting On in the World," page 305).

A word to the young men who are, or intend to become farmers : If you have chosen farming as a profession, I would advise you to graduate at some of our Agricultural Colleges, before entering upon your life work. I mention this, as it is a daily regret with myself, that I left school at a very early age, and that I know so little of the science of Agriculture, as taught at the present day in nearly every State. If you are to be a farmer, you cannot know too much of what pertains to the most approved and scientific methods.

You have not the virgin soil that your fathers had, on which to practice farming, but soil, which they, through ignorance, have exhausted, and which will now require your utmost skill to redeem.

Anyone can exhaust the soil, but no one but a wise man can win it back to its former state of productiveness.

Ever remembering that your plants, like your animals, live, feed, grow and die, and, that by feeding them alike plentifully, they will produce bountifully.

In this respect, it is the liberal hand that maketh rich.

INDEX.

SUMMER SOILING.

	PAGE.
Advantages of Soiling	38
1—Saving of Land,	38
2—Saving of Fences,	41
3—Saving of Food,	33
4—Better Condition and Comfort of Stock,	44
5—Greater production of Beef, Butter or Milk,	48
6—Increased quantity and quality of Manure,	52
7—Increased productiveness of the soil,	54
Commercial Fertilizers,	29
Cost of.	30
Compared with Stable Manure,	30, 31
Comparative values of Grains and Forage as Animal and Plant Food, (tables,)	20, 22, 23, 24
Correspondence from Charles Barton,	104
" " Charles Francis,	100
Conclusion,	147
Cows,	16
Absurd Theories,	17
As a machine,	1, 13
Cost of keeping, (table,)	18, 19
Feeding to produce beef, milk or butter,	14, 15, 16
Crops for Soiling Cattle,	38
Barley,	62
Common Millet,	69
Clover,	66
Corn,	63
Hungarian Grass (Millet,)	69
Lucern,	67
Oats, or Oats and Peas,	65
Rye,	61
Crops for Soiling Sheep,	96
Rape,	97
Tares (Vetches,)	96
Turnips,	98

	PAGE.
Cutting Soiling Crops,	81
Reaper for,	81
Drawing green fodder,	82
Wagon for,	82
Education for Farmers,	144
Feeding Soiling Crops (cattle),	83
Caution in feeding,	83
Racks for feeding,	83
Objections to racks,	60
When and how often to feed, (example),	84
Feeding Soiling Crops (sheep,)	83
Lamb creep,	97, 102, 107
Moveable racks,	100
Rotation of crops,	100
When and how to feed,	102
Feeding Shed for sheep (illustration,)	107
Advantages of,	107
Construction and location,	106, 107
Rotation system,	108
Hurdling and Folding (sheep,)	102
Construction of hurdles,	103
Portable fencing,	104
Rope and Wire Netting (cost of),	104
Horses—Soiling,	85, 90
Advantages,	91
Brood Mares and Colts,	90
Influence of food,	7
Economy in feeding,	12
Improvement of farm stock,	9
Origin of different breeds,	8
Reproductive powers of animals,	11
Manure,	24
Barnyard,	25
Green Crops as	25
Liquid Manure,	26
Absorbents,	88
Application,	28
Compared with solid, (table,)	27
Saving,	27
Value,	27
Objections to Soiling,	57
Extra Labor,	57
Want of exercise,	59
Permanent Pasture,	105
Variety of Grasses for,	105

	PAGE.
Rotation of Soiling Crops,	75
Amount of Land required,	76
How to begin,	76
Plowing and fitting the ground,	77
Roots,	71
Beets,	71
Cabbage,	73
Kohl Rabi,	74
Sheep Soiling,	92
Advantages,	93, 94
Soiling,	35
Why adopted,	35
Stables,	87
Construction,	134–135
Fastnings for cows,	88
Floor for cows,	87
Ventilation of…	86
System,…	137

WINTER SOILING.

	PAGE.
Advantages,…	114
Cutting the fodder,	125
Reaper for	125
Drawing,	125
Loading and Unloading,	126
Feeding Ensilage,	128
Compared with hay (table),	132
Cost of Ensilage (table),	130
Experiment in Feeding,	129
Growing crops for Ensilage,	123
Amount of seed per acre,	125
Method of culture,…	124
Number of tons per acre,	123
Opening Silo,…	127
Pressing or weighting,	126
Silos,…	120
How to build,…	120
Where to build,…	121, 135
Soiling Ensilage,	136
Table showing comparative value of Pasturing, Ensilage and Soiling,…	137

D. M. OSBORNE & CO.,

MANUFACTURERS OF

Mowing and Reaping Machines,

AUBURN, N. Y.

THE INDEPENDENT OSBORNE REAPER.

The No. 3 Reaper has been an acknowledged leader in Reaping Machines for very many years. It has many imitators, but no equals outside of its own family. It cuts a wider swath than any other self-raking reaper, and excels also in its capacity for work. It has a strong wrought iron frame with a flexible bar and independent motion in the axle-plate, giving it great strength and perfect adjustment, so that the severest strains in work, do not affect its capacity. It is a Front-cut reaper, with the driver's seat far enough in the rear to give him perfect oversight of the working of the machine. The driver's weight balances the machine, and there is no weight upon the necks of the team, nor any side draft.

The Rolling-head Rake is used, and is, undoubtedly, the most perfect form of self-rake now made. The four rakes can be made to act either as rakes or beaters—any one of them can be used to rake off, while the other three reel on the grain. Any number of the rakes, from one to all four, may be used to rake off or any one or more, may be used to work automatically, without any attention from the driver.

It is admirable for saving lodged and tangled grain, the forward part of the machine being depressed so that the guards run close to the ground, and the rakes following, can reach the worst lodged and flattened grain and bring it to the knives.

It is very light, easily handled and thoroughly durable; well adapted for any kind of service—wet or dry, rough or smooth.

The No. 3 Osborne Reaper has been widely and successfully used for cutting all kinds of Green Fodder, for use as Ensilage.

It handles Green Sowed Corn in a very admirable and satisfactory manner.

ENSILAGE.

STOCK RAISERS, DAIRYMEN, Etc.

ROSS GIANT AND LITTLE GIANT CUTTERS.

"SPECIAL."

These Cutters are very HEAVY, STRONG and DURABLE; are made from the best of material and best of workmanship, and are what they are intended to be, the very BEST Cutters it is possible to build.

They are expressly intended for Ensilage, and Stock Raisers etc., wishing a good machine.

They are fitted with our Patent Safety Fly Wheels, Extensible Joints, Universal Action Feeding Rollers, Convex-faced Gears, Quick Fastenings. Ring Rollers, &c., which relieve the cutters from all strain, and prevents breakages, and insures safety to operators. Can be driven upon either side.

The machines cut from 2 to 4 times as rapidly as any other make; require little power to drive, and leaves the fodder in splendid condition.

They are used by the largest Ensilagists in the United States. Send for Circular, Price Lists and Testimonials.

E. W. ROSS & CO.,
FULTON, N. Y.

AGRICULTURAL REVIEW
and JOURNAL of the
American Agricultural Association.
Published Quarterly,
(January, April, July and October.)

The *Review* is a Magazine of about 200 pages, printed in the very best manner. Agriculture, Commerce, Transportation, and kindred subjects, are discussed by the most scientific and practical men of the country. The circulation has increased at a most rapid rate, proving that such a publication was needed. Each Number is a volume in itself, and worth the price of a year's subscription.

Among others who have contributed, are Prof. W. O. Atwater, Middletown, Conn., Dr. E. Lewis, Sturtevant, N. Y., Experimental Station, Prof. Arthur L. Yerry, Williamstown, Mass., J. B. Lawless, *L. L. D.*, Rothamstead, England, Rear Admiral Daniel Ammen, Washington, D. C., Hon. Edward Atkinson, Boston, Mass., Prof. J. P. Roberts, Ithaca, N. Y., Dr. Byron D. Halstead, Am. Agriculturist, Willis P. Hazard, Westchester, Pa., Prof. J. M. McBride, Knoxville, Tenn.

Terms $3.00 for this year, including membership in the American Agricultural Association.

Single copies 50 cents.

The American Agricultural Association having decided to hold a Grand Exposition of Agricultural Products, the Review and Journal will give the latest and fullest information thereto, making it especially *interesting* and *valuable*.

Subscriptions should be sent to

JOSEPH H. TEALL, *Editor and Publisher*,

26 UNIVERSITY PLACE, - NEW YORK.

CORN PLANTING FOR ENSILAGE.

Among the factors most necessary for the profitable production of Ensilage, none command more earnest consideration than the planting of the crop from which it is to be made. The first element, and lacking which, others are valueless, is the ability to plant the Corn in the best possible shape, insuring most abundant and rapid growth. This is unquestionably done by planting in drills.

At the time of planting, putting in the drills Fertilizers, which will add largely to the natural fertility of the soil. After this, the arranging of proper distance between the drills is almost as necessary to secure pronounced success.

Both of these results are readily attained by the use of the FARMERS' FAVORITE GRAIN DRILL.

We shall be pleased to answer all inquiries with reference to this implement, furnish Circulars, and when desired, furnish these Drills to Farmers, under fullest guarantee of their superiority.

This Drill is Eminently Superior to all others, for ordinary field work, as well as field planting of Corn. Address

BICKFORD & HOFFMAN,
Patentees and Manufacturers,   MACEDON, Wayne Co., N. Y

www.ingramcontent.com/pod-product-compliance
Lightning Source LLC
Chambersburg PA
CBHW030311170426
43202CB00009B/967